Tetsuya Hoya

Artificial Mind System – Kernel Memory Approach

Studies in Computational Intelligence, Volume 1

Editor-in-chief
Prof. Janusz Kacprzyk
Systems Research Institute
Polish Academy of Sciences
ul. Newelska 6
01-447 Warsaw
Poland
E-mail: kacprzyk@ibspan.waw.pl

Further volumes of this series
can be found on our homepage:
springeronline.com

Vol. 1. Tetsuya Hoya
*Artificial Mind System – Kernel Memory
Approach,* 2005
ISBN 3-540-26072-2

Tetsuya Hoya

Artificial Mind System

Kernel Memory Approach

 Springer

Dr. Tetsuya Hoya
RIKEN Brain Science Institute
Laboratory for Advanced
Brain Signal Processing
2-1 Hirosawa, Wako-Shi
Saitama, 351-0198
Japan
E-mail: hoya@brain.riken.jp

ISSN print edition: 1860-949X
ISSN electronic edition: 1860-9503
ISBN-10 3-642-42472-4 Springer Berlin Heidelberg New York
ISBN-13 978-3-642-42472-4 Springer Berlin Heidelberg New York

Springer is a part of Springer Science+Business Media
springeronline.com
© Springer-Verlag Berlin Heidelberg 2005
Softcover re-print of the Hardcover 1st edition 2005

Typesetting: by the authors and TechBooks using a Springer LaTeX macro package

Printed on acid-free paper   SPIN: 10997444   89/TechBooks   5 4 3 2 1 0

To my colleagues, educators, and my family

Preface

This book was written from an engineer's perspective of mind. So far, although quite a large amount of literature on the topic of the mind has appeared from various disciplines; in this research monograph, I have tried to draw a picture of the holistic model of an artificial mind system and its behaviour, as concretely as possible, within a unified context, which could eventually lead to practical realisation in terms of hardware or software. With a view that *"mind is a system always evolving"*, ideas inspired/motivated from many branches of studies related to brain science are integrated within the text, i.e. artificial intelligence, cognitive science/psychology, connectionism, consciousness studies, general neuroscience, linguistics, pattern recognition/data clustering, robotics, and signal processing. The intention is then to expose the reader to a broad spectrum of interesting areas in general brain science/mind-oriented studies.

I decided to write this monograph partly because now I think is the right time to reflect at what stage we currently are and then where we should go towards the development of "brain-style" computers, which is counted as one of the major directions conducted by the group of "creating the brain" within the brain science institute, RIKEN.

Although I have done my best, I admit that for some parts of the holistic model only the frameworks are given and the descriptions may be deemed to be insufficient. However, I am inclined to say that such parts must be heavily dependent upon specific purposes and should be developed with careful consideration during the domain-related design process (see also the Statements to be given next), which is likely to require material outside of the scope of this book.

Moreover, it is sometimes a matter of dispute whether a proposed approach/model is biologically plausible or not. However, my stance, as an engineer, is that, although it may be sometimes useful to understand the underlying principles and then exploit them for the development of the "artificial" mind system, only digging into such a dispute will not be so beneficial for the development, once we set our ultimate goal to construct the mechanisms functioning akin to the brain/mind. (Imagine how fruitless it is to argue, for instance, only about the biological plausibility of an airplane; an artificial object that can fly, but not like a bird.) Hence, the primary objective of this monograph is not to seek such a plausible model but rather to provide a basis for imitating the functionalities.

On the other hand, it seems that the current trend in general connectionism rather focuses upon more and more sophisticated learning mechanisms or their highly-mathematical justifications without showing a clear direction/evidence of how these are related to imitating such functionalities of brain/mind, which many times brought me a simple question, *"Do we really need to rely on such highly complex tools, for the pursuit of creating the virtual brain/mind?"* This was also a good reason to decide writing the book.

Nevertheless, I hope that the reader enjoys reading it and believe that this monograph will give some new research opportunities, ideas, and further insights in the study of artificial intelligence, connectionism, and the mind. Then, I believe that the book will provide a ground for the scientific communications amongst various relevant disciplines.

Acknowledgment

First of all, I am deeply indebted to Professor Andrzej Cichocki, Head of the Laboratory for Advanced Brain Signal Processing, Brain Science Institute (BSI), the Institute of Physical and Chemical Research (RIKEN), who is on leave from Warsaw Institute of Technology and gave me a wonderful opportunity to work with the colleagues at BSI. He is one of the mentors as well as the supervisors of my research activities, since I joined the laboratory in Oct. 2000, and kindly allowed me to spend time writing this monograph. Without his continuous encouragement and support, this work would never have been completed. The book is moreover the outcome of the incessant excitement and stimulation gained over the last few years from the congenial atmosphere within the laboratory at BSI-RIKEN. Therefore, my sincere gratitude goes to Professor Shun-Ichi Amari, the director, and Professor Masao Ito, the former director of BSI-RIKEN whose international standing and profound knowledge gained from various brain science-oriented studies have coalized at BSI-RIKEN, where exciting research activities have been conducted by maximally exploiting the centre's marvelous facilities since its foundation in 1997. I am much indebted to Professor Jonathon Chambers, Cardiff Professorial Fellow of Digital Signal Processing, Cardiff School of Engineering, Cardiff University, who was my former supervisor during my post-doc period from Sept. 1997 to Aug. 2000, at the Department of Electrical and Electronic Engineering, Imperial College of Science, Technology, and Medicine, University of London, for undertaking the laborious proofreading of the entire book written by a non-native English speaker. Remembering the exciting days in London, I would like to express my gratitude to Professor Anthony G.

Constantinides of Imperial College London, who was the supervisor for my Ph.D. thesis and gave me excellent direction and inspiration. Many thanks also go to my colleagues in BSI, collaborators, and many visitors to the ABSP laboratory, especially Dr. Danilo P. Mandic at Imperial College London, who has continuously encouraged me in various ways for this monograph writing, Professor Hajime Asama, the University of Tokyo, Professor Michio Sugeno, the former Head of the Laboratory for Language-Based Intelligent Systems, BSI-RIKEN, Dr. Chie Nakatani and Professor Cees V. Leeuwen of the Laboratory for Perceptual Dynamics, BSI-RIKEN, Professor Jianting Cao of the Saitama Institute of Technology, Dr. Shuxue Ding, at the University of Aizu, Professor Allan K. Barros, at the University of Maranhão (UFMA), and the students within the group headed by Professor Yoshihisa Ishida, who was my former supervisor during my master's period, at the Department of Electronics and Communication, School of Science and Engineering, Meiji University, for their advice, fruitful discussions, inspirations, and useful comments.

Finally, I must acknowledge the continuous and invaluable help and encouragement of my family and many of my friends during the monograph writing.

BSI-RIKEN, Saitama
April 2005
Tetsuya Hoya

Statements

Before moving ahead to the contents of the research monograph, there is one thing to always bear in our mind and then we need to ask ourselves from time to time, "What if we successfully developed artificial intelligence (AI) or humanoids that behaves as real mind/humans? Is it really beneficial to human-kind and also to other species?" In the middle of the last century, the country Japan unfortunately became a single (and hopefully the last) country in the world history that actually experienced the aftermath of nuclear bombs. Then, only a few years later into the new millennium (2000), we are frequently made aware of the peril of bio-hazard, resulting from the advancement in biology and genetics, as well as the world-wide environmental problems. The same could potentially happen if we succeeded the development and thereby exploited recklessly the intelligent mechanisms functioning quite akin to creatures/humans and eventually may lead to our existence being endangered in the long run. In 1951, the cartoonist Osamu Tezuka gave birth to the astroboy named "Atom" in his works. Now, his cartoons do not remain as a mere fiction but are like to become reality in the near future. Then, they warn us how our life can be dramatically changed by having such intelligent robots within our society; as a summary, in the future we may face to the relevant issues as raised by Russell and Norvig (2003):

- People might lose their jobs to automation;
- People might have too much (or too little) leisure time;
- People might lose their sense of being unique;
- People might lose some of their privacy rights;
- The use of AI systems might result in a loss of accountability;
- The success of AI might mean the end of the human race.

In a similar context, the well-known novel "Frankenstein" (1818) by Mary Shelley also predicted such a day to come. These works, therefore, strongly suggest that it is high time we really needed to start contemplating the (near) future, where AIs or robots are ubiquitous in the surrounding environment, what we humans are in such a situation, and what sort of actions are necessary to be taken by us. I thus hope that the reader also takes these emerging issues very seriously and proceeds to the contents of the book.

Contents

List of Abbreviations

ADF	ADaptive Filter
AI	Artificial Intelligence
ALCOVE	Attention Learning COVEring map
ALE	Adaptive Line Enhancer
AMS	Artificial Mind System
ANN	Artificial Neural Network
ARTMAP	Adative Resonance Theory MAP
ASE	Adaptive Signal Enhancer
BP	Back-Propagation
BSE	Blind Signal Extraction
BSP	Blind Signal Processing
BSS	Blind Source Separation
CMOS	Complimentary Metal-Oxide Semiconductor
CR	Conditioned Response
CS	Conditioned Stimuli
DASE	Dual Adaptive Signal Enhancer
DFT	Discrete Fourier Transform
DOA	Direction Of Arrival
ECG	ElectroCardioGraphy
EEG	ElectroEncephaloGraphy
EGO	Emotionally GrOunded
EMG	ElectroMyoGraphy
EVD	EigenValue Decomposition
FIR	Finite Impulse Response
FFT	Fast Fourier Transform
fMRI	functional Magnetic Resonance Imaging
GCM	Generalised Context Model
GMM	Gaussian Mixture Model
GRNN	Generalised Regression Neural Network
HA-GRNN	Hierarchically Arranged Generalised Regression
HMM	Hidden Markov Model

HRNN	Hopfield-type Recurrent Neural Network
ICA	Independent Component Analysis
i.i.d.	Independent Identically Distributed
KF	Kernel Function
KM	Kernel Memory
K-Line	Knowledge-Line
LAD	Language Acquisition Device
LIFO	Last-In-Fast-Out
LMS	Least Mean Square
LPC	Linear Predictive Coding
LTD	Long Term Depression
LTM	Long Term Memory
MDIMO	Multi-Domain Input Multi-Output
MEG	MagnetoEncephaloGraphy
MIMO	Multi-Input Multi-Output
MLP-NN	Multi-Layered Perceptron Neural Network
MORSEL	Multiple Object Recognition and Attentional Selection
M-SSP	Multi-stage Sliding Subspace Projection
NLMS	Normalised Least Mean Square
NM	Neural Memory
NN	Neural Network
NR	Noise Reduction
NSS	Nonlinear Spectral Subtraction
PET	Positron Emission Tomography
PNN	Probabilistic Neural Network
PRS	Perceptual Representation System
PSD	Power Spectral Density
QMF	Quadrature Mirror Filter
RBF	Radial Basis Function
SAD	Sound Activity Detection
SAIM	Selective Attention for Identification Model
SDIMO	Single-Domain-Input Single Output
SE	Signal Separation
SFS	Speech Filing System
SIMO	Single-Input Single Output
SLAM	SeLective Attention Model
SNR	Signal-to-Noise Ratio
SOBI	Second-Order Blind Identification
SOFM	Self-Organising Feature Map
SOKM	Self-Organising Kernel Memory
SPECT	Single-Photon Emission Computed Tomography
SRN	Simple Recurrent Network
SS	Signal Separation
SSP	Sliding Subspace Projection
STM	Short Term Memory

SVD	Singular Value Decomposition
SVM	Support Vector Machine
TDNN	Time Delay Neural Network
UR	Unconditioned Response
US	Unconditioned Stimuli
XOR	eXclusive OR

1

Introduction

1.1 Mind, Brain, and Artificial Interpretation

"What is mind?" When you are asked such a question, you may be probably confused, because you do not exactly know how to answer, though you
frequently use the word "mind" in daily conversation to describe your conditions, experiences, feelings, mental states, and so on. On the other hand,
many people have so far tackled the topic of how science can handle the mind
and its operation.

This monograph is an attempt to deal with the topic of the mind from
the perspective of certain engineering principles, i.e. connectionism and signal
processing studies, whilst weaving a view from cognitive science/psychological
studies (see Gazzaniga et al., 2002) as the supporting background. Hence, as
in the title of the book, the objective of this monograph is primarily to pro-
pose a direction/scope of how an "artificial" mind system can be developed,
based upon these disciplines. Therefore, by the term "artificial", the aim is
ultimately to develop a mechanical system that imitates the various functionalities of the mind and is implemented within intelligent robots (thus, the aim
is also relevant to the general purpose of "creating the brain").

As current mind research is heavily indebted to the dramatic progress in
brain science, in which the brain, a natural being so elaborately organised, as
a consequence of thousands-and-thousands of years of natural evolution, has
been treated as a physical substance and studied by analysing the functionalities of the tissues therein. Brain science has therefore been established with
the support of rapid advancement in measurement technology and thereby
yielded better understanding of how the brain works.

The history of mind/brain research backdates to the Aristotle period of
time (i.e. 384–322 B.C.), a Greek philosopher and scientist who first formulated a precise set of laws governing the rational part of the mind, followed
by the birth of philosophy (i.e. 428 B.C.), and then by that of mathematics
(c.800), economics (1776), neuroscience (1861), psychology (1879), computer

Tetsuya Hoya: *Artificial Mind System – Kernel Memory Approach*, Studies in Computational
Intelligence (SCI) **1**, 1–8 (2005)
www.springerlink.com

engineering (1940), control theory and cybernetics (1948), artificial intelligence (AI) and cognitive science (1956), and linguistics (1957) (for a concise summary, see also Russell and Norvig, 2003), all the disciplines of which are somewhat relevant to the studies of mind (cf. e.g. Fodor, 1983; Minsky, 1985; Grossberg, 1988; Dennett, 1988; Edelman, 1992; Anderson, 1993; Crane, 1995; Greenfield, 1995; Aleksander, 1996; Kawato, 1996; Chalmers, 1996; Kitamura, 2000; Pfeifer and Scheier, 2000; McDermott, 2001; Shibata, 2001). This stream has led to the recent development of robots which imitate the behaviours of creatures, or humanoids (albeit still primitive), especially those realised by several Japanese industries.

In the philosophical context, the topic of the mind has alternatively been treated as the so-called mind-brain problem, as Descartes (1596-1650) once gave a clear distinction between mind and body (brain), ontology, or within the context of *consciousness* (cf. e.g. Turing, 1950; Terasawa, 1984; Dennett, 1988; Searle, 1992; Greenfield, 1995; Aleksander, 1996; Chalmers, 1996; Osaka, 1997; Pinker, 1997; Hobson, 1999; Shimojo, 1999; Gazzaniga et al., 2002). Then, there are, roughly speaking, two well-known philosophical standpoints to start discussing the issue of mind – *dualism* and *materialism*; *Dualism*, as supported by the philosophers such as Descartes and Wittgenstein, is a standpoint that, unlike animals, the human mind exists by its own and hence must be separated from the physical substance of the body/brain, whilst the opponent *materialism* holds the notion that the mind is nothing more than the phenomenon of the processing occurring within the brain. Hence, the book is written, generally within the latter principle.

1.2 Multi-Disciplinary Nature of the Research

Figure 1.1 shows the author's scope of active studies In the area and their mutual relationships for the necessity of "creating the brain"; it is considered

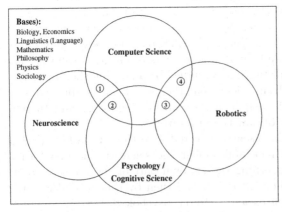

Bases):
Biology, Economics
Linguistics (Language)
Mathematics
Philosophy
Physics
Sociology

Computer Science

Neuroscience

Robotics

Psychology /
Cognitive Science

4 Major Composite Groups):

① Biophysics

② Animal / Developmental Studies ;
   Measurement Studies – EEG /
   MEG / fMRI / PET / SPECT, etc. ;

③ Connectionism ;
   Consciousness Studies (partially
   relevant to Neuroscience)

④ Artificial Intelligence ;
   Control Theory ;
   Optimisation Theory ;
   Signal Processing ;
   Statistics

(In the above, connectionism lies loosely
across all the four fundamentals.)

Fig. 1.1. Creating the brain – a multi-disciplinary area of research

that the direction towards "creating the brain" consists of (at least) the 12 core studies/scientific bases and other 11 inter-related subjects which respectively fall in to the four major composite groups. Thus, within the author's scope, a total of (but not limited to) 23 areas of the studies are simultaneously taken into account for the pursuit of this challenging topic – i.e. 1) animal studies, 2) artificial intelligence, 3) biology, 4) biophysics, 5) (general) cognitive science, 6) computer science, 7) connectionism (or, more conventionally, artificial neural networks), 8) consciousness studies, 9) control theory, 10) developmental studies, 11) economics, 12) linguistics (language), 13) mathematics (in general), 14) measurement studies relevant to brain waves – such as electroencephalography (EEG), magnetoencephalography (MEG), functional magnetic resonance imaging (fMRI), positron-emission tomography (PET), or single photon emission computed tomography (SPECT) – 15) neuroscience, 16) optimisation theory, 17) philosophy, 18) physics, 19) (various branches of) psychology, 20) robotics, 21) signal processing, 22) sociology, and finally 23) statistics, all of which are, needless to say, currently quite active areas of research. It is then considered that the seventh study, i.e. connectionism, lies (loosely) across all the fundamental studies, i.e. computer science, neuroscience, cognitive science/psychology, and robotics.

In other words, the topic must be essentially based upon a multi-disciplinary nature of research. Therefore, to achieve the ultimate goal, it is inevitable that we do not bury ourselves in a single narrow area of research but always bear in our mind the global picture as well as the cross-fertilisation of the research activities.

1.3 The Stance to Conquest the Intellectual Giant

Although it is highly attractive to progress the research of "creating the brain", as stated earlier (in the Statements), we should always be rather careful about further advancing the activity in "creating the brain" (since it may eventually lead to endanger the existence of ourselves).

Then, here, let us limit the necessity of "creating the brain" to the purpose of *"creating the artificial system that behaves or functions as the mind"*, or simply, *"create the virtual mind"*, since, if we denote "creating the brain", it may also imply to develop totally biologically feasible models of brain, the topic of which has to be extremely carefully treated (see the Statements) and hence is beyond the scope of this book.

Therefore, the following four major phases should be embraced in order to conduct the research activities within the context of "creating the virtual mind":

Phase 1) Observe the "phenomena" of real brains, by maximally exploiting the currently available brain-wave measurements (This is hence rather relevant to the issues of *"understanding the brain"*), and the activities of real life (i.e. not limited to humans), as carefully as possible. (Needless to say, it is also fundamentally significant to advance such measurement technology, in parallel with this phase.)

Phase 2) Model the brain activities/phenomena, by means of engineering tools and develop the feasible as well as unified concepts, supported by the principles from the four core subjects – 1) computer science, 2) neuroscience, 3) cognitive science/psychology, and 4) robotics.

Phase 3) Realise the models in terms of hardware or software (or, even, the so-called "wetware", though as aforementioned, this must also be carefully dealt within the context of humanity or scientific philosophy) and validate if they actually imitate the behaviour of the brain/mind.

Phase 4) Investigate the results obtained in the third phase amongst the multiple disciplines (23 in total) given earlier. Return to the first phase.

Note that, in the above, it is not meant that the four phases should always be subsequent but rather suggested that the inter-phase activities also be encouraged.

Hence, the purpose of this book is generally to provide the accounts relevant to both Phases 2) and 3) above.

1.4 The Artificial Mind System Based Upon Kernel Memory Concept

The concept of the artificial mind system was originally inspired by the so-called "modularity of mind" principle (Fodor, 1983; Hobson, 1999), i.e. the functionality of the mind is subdivided into the respective modules, each of which is responsible for a particular psychological function. (However, note that here the "module" is not always referred to as merely a distinct "agent", as often appeared in the reductionist context.)

Hobson (Hobson, 1999) proposed that consciousness consists of the constituents as tabulated in Table 1.1 (then, it is considered that each constituent also corresponds to the notion of "module" within the modularity principle of mind Fodor (1983)). As in the table, the constituents can be subdivided into three major groups, i.e. i) input sources, ii) assimilating processing, and iii) output actions.

Therefore, with the supportive studies by Fodor (Fodor, 1983) and Hobson (Hobson, 1999), the artificial system imitating the various functionalities of mind can macroscopically be regarded as an input-output system and developed based upon the modularity principle. Then, the objective here is to model the respective constituents of mind similar to those in Table 1.1 and their mutual data processing within the engineering context (i.e. realised in terms of hardware/software).

Table 1.1. Constituents of consciousness (adapted from Hobson, 1999)

Input Sources	
Sensation	Receival of input data
Perception	Representation of input data
Attention	Selection of input data
Emotion	Emotion of the representation
Instinct	Innate tendency of the actions
Assimilating Processes	
Memory	Recall of cumurated evocation
Thinking	Response to the evocation
Language	Symbolisation of the evocation
Intention	Evocation of aim
Orientation	Evocation of time, place, and person
Learning	Automatic recording of experience
Output Actions	
Intentional Behaviour	Decision making
Motion	Actions and motions

On the other hand, it still seems that the progress in connectionism has not reached a sufficient level to explain/model the higher-order functionalities of brain/mind; the current issues, e.g. appeared in many journal/conference papers, in the field of artificial neural networks (ANNs) are mostly concentrated around development of more sophisticated algorithms, the performance improvement versus the existing models, mostly discussed within the same problem formulation, or the mathematical analysis/justification of the behaviours of the models proposed so far (see also e.g. Stork, 1989; Roy, 2000), without showing a clear/further direction of how these works are related to answer one of the most fundamentally important problems: how the various functionalities relevant to the real brain/mind can be represented by such models. This has unfortunately detracted much interest in exploiting the current ANN models for explaining higher functions of the brain/mind. Moreover, Herbert Simon, the Nobel prize winner in economics (in 1978), also implied (Simon, 1996) that it is not always necessary to imitate the functionality from the microscopic level for such a highly complex organisation as the brain. Then, by following this principle, the kernel memory concept, which will appear in the first part of this monograph, is here given to (hopefully) cope with the stalling situation.

The kernel memory is based upon a simple element called the *kernel unit*, which can internally hold [a chunk of] data (thus representing "memory"; stored in the form of *template data*) and then (essentially) does the pattern matching between the input and template data, using the similarity measurement given as its *kernel function*, and its connection(s) to other units. Then, unlike ordinary ANN models (for a survey, see Haykin, 1994), the connections simply represent the *strengths* between the respective kernel units in order to propagate the activation(s) of the corresponding kernel units, and the update of the weight values on such connections does not resort to any gradient-descent type algorithm, whilst holding a number of attractive properties. Hence, it may also be seen that kernel memory concept can replace conventional symbol-grounding connectionist models.

In the second part of the book, it will be described how the kernel memory concept is incorporated into the formation of each module within the artificial mind system (AMS).

1.5 The Organisation of the Book

As aforementioned, this book is divided into two parts: the first part, i.e. from Chap. 2 to 4, provides the neural foundation for the development of the AMS and the modules within it, as well as their mutual data processing, to be described in detail in the second part, i.e. from Chap. 5 to 11.

In the following Chap. 2, we briefly review the conventional ANN models, such as the associative memory, Hopfield's recurrent neural networks (HRNNs) (Hopfield, 1982), multi-layered perceptron neural networks (MLP-NNs), which are normally trained using the so-called back-propagation (BP) algorithm (Amari, 1967; Bryson and Ho, 1969; Werbos, 1974; Parker, 1985; Rumelhart et al., 1986), self-organising feature maps (SOFMs) (Kohonen, 1997), and a variant of radial basis function neural networks (RBF-NNs) (Broomhead and Lowe, 1988; Moody and Darken, 1989; Renals, 1989; Poggio and Girosi, 1990) (for a concise survey of the ANN models, see also Haykin, 1994). Then, amongst a family of RBF-NNs, we highlight the two models, i.e. probabilistic neural networks (PNNs) (Specht, 1988, 1990) and generalised regression neural networks (GRNNs) (Specht, 1991), and investigate the useful properties of these two models.

Chapter 3 gives a basis for a new paradigm of the connectionist model, namely, the kernel memory concept, which can also be seen as the generalisation of PNNs/GRNNs, followed by the description of the novel self-organising kernel memory (SOKM) model in Chap. 4. The weight updating (or learning) rule for SOKMs is motivated from the original Hebbian postulate between a pair of cells (Hebb, 1949). In both Chaps. 3 and 4, it will be described that the kernel memory (KM) not only inherits the attractive properties of PNNs/GRNNs but also can be exploited to establish the neural basis for modelling the various functionalities of the mind, which will be extensively described in the rest of the book.

The opening chapter for the second part firstly proposes a holistic model of the AMS (i.e. in Chap. 5) and discusses how it is organised within the principle of modularity of the mind (Fodor, 1983; Hobson, 1999) and the functionality of each constituent (i.e. module), through a descriptive example. It is hence considered that the AMS is composed of a total of 14 modules; one single input, i.e. the input: sensation module, two output modules, i.e. the primary and secondary (perceptual) outputs, and remaining 11 modules, each of which represents the corresponding cognitive/psychological function: 1) attention, 2) emotion, 3,4) explicit/implicit long-term memory (LTM), 5) instinct: innate structure, 6), intention, 7) intuition, 8) language, 9) semantic networks/lexicon, 10) short-term memory (STM)/working memory, and 11) thinking module, and their interactions. Then, the subsequent Chaps. 6–10 are devoted to the description of the respective modules in detail.

In Chap. 6, the sensation module of the AMS is considered as the module responsible for the sensory inputs arriving at the AMS and represented by a cascade of pre-processing units, e.g. the units performing sound activity detection (SAD), noise reduction (NR), or signal extraction (SE)/separation (SS), all of which are active areas of study in signal processing. Then, as a practical example, we consider the problem of noise reduction for stereophonic speech signals with an extensive simulation study. Although the noise reduction model to be described is totally based upon a signal processing approach, it is thought that the model can be incorporated as a practical noise reduction part of the mechanism within the sensation module of the AMS. Hence, it is expected that, for the material in Sect. 6.2.2, as well as for the blind speech extraction model described in Sect. 8.5, the reader is familiar with signal processing and thus has the necessary background in linear algebra theory. Next, within the AMS context, the perception is simply defined as pattern recognition by accessing the memory contents of the LTM-oriented modules and treated as the secondary output.

Chapter 7 deals rather in depth with the notion of learning and discusses the relevant issues, such as supervised/unsupervised learning and target responses (or interchangeably the "teachers" signals), all of which invariably appear in ordinary connectionism, within the AMS context. Then, an example of a combined self-evolutionary feature extraction and pattern recognition is considered based upon the model of SOKM in Chap. 4.

Subsequently, in Chap. 8, the memory modules within the AMS, i.e. both the explicit and implicit LTM, STM/working memory, and the other two LTM-oriented modules – semantic networks/lexicon and instinct: innate structure modules – are described in detail in terms of the kernel memory principle. Then, we consider a speech extraction system, as well as its extension to convolutive mixtures, based upon a combined subband independent component analysis (ICA) and neural memory as the embodiment of both the sensation and LTM modules.

Chapter 9 focuses upon the two memory-oriented modules of language and thinking, followed by interpreting the abstract notions related to mind within the AMS context in Chap. 10. In Chap. 10, the four psychological function-oriented modules within the AMS, i.e. attention, emotion, intention, and intuition, will be described, all based upon the kernel memory concept. In the later part of Chap. 10, we also consider how the four modules of attention, intuition, LTM, and STM/working memory can be embodied and incorporated to construct an intelligent pattern recognition system, through a simulation study. Then, the extended model that implements both the notions of emotion and procedural memory is considered.

In Chap. 11, with a brief summary of the modules, we will outline the enigmatic issue of consciousness within the AMS context, followed by the provision of a short note on the brain mechanism for intelligent robots. Then, the book is concluded with a comprehensive bibliography.

The Neural Foundations

2

From Classical Connectionist Models to Probabilistic/Generalised Regression Neural Networks (PNNs/GRNNs)

2.1 Perspective

This chapter begins by briefly summarising some of the well-known classical connectionist/artificial neural network models such as multi-layered perceptron neural networks (MLP-NNs), radial basis function neural networks (RBF-NNs), self-organising feature maps (SOFMs), associative memory, and Hopfield-type recurrent neural networks (HRNNs). These models are shown to normally require iterative and/or complex parameter approximation procedures, and it is highlighted why these approaches have in general lost interest in modelling the psychological functions and developing artificial intelligence (in a more realistic sense).

Probabilistic neural networks (PNNs) (Specht, 1988) and generalised regression neural networks (GRNNs) (Specht, 1991) are discussed next. These two networks are often regarded as variants of RBF-NNs (Broomhead and Lowe, 1988; Moody and Darken, 1989; Renals, 1989; Poggio and Girosi, 1990), but, unlike ordinary RBF-NNs, have several inherent and useful properties, i.e. 1) straightforward network configuration (Hoya and Chambers, 2001a; Hoya, 2004b), 2) robust classification performance, and 3) capability in accommodating new classes (Hoya, 2003a).

These properties are not only desirable for on-line data processing but also inevitable for modelling psychological functions (Hoya, 2004b), which eventually leads to the development of kernel memory concept to be described in the subsequent chapters.

Finally, to emphasise the attractive properties of PNNs/GRNNs, a more informative description by means of the comparison with some common connectionist models and PNNs/GRNNs is given.

Tetsuya Hoya: *Artificial Mind System – Kernel Memory Approach*, Studies in Computational Intelligence (SCI) **1**, 11–29 (2005)
www.springerlink.com

2.2 Classical Connectionist/Artificial Neural Network Models

In the last few decades, the rapid advancements of computer technology have enabled studies in artificial neural networks or, in a more general terminology, *connectionism*, to flourish. Utility in various real world situations has been demonstrated, whilst the theoretical aspects of the studies had been provided long before the period.

2.2.1 Multi-Layered Perceptron/Radial Basis Function Neural Networks, and Self-Organising Feature Maps

In the artificial neural network field, multi-layered perceptron neural networks (MLP-NNs), which were pioneered around the early 1960's (Rosenblatt, 1958, 1962; Widrow, 1962), have played a central role in pattern recognition tasks (Bishop, 1996). In MLP-NNs, sigmoidal (or, often colloquially termed "squash", from the shape of the envelope) functions are used for the nonlinearity, and the network parameters, such as the weight vectors between the input and hidden layers and those between hidden and output layers, are usually adjusted by the back-propagation (BP) algorithm (Amari (1967); Bryson and Ho (1969); Werbos (1974); Parker (1985); Rumelhart et al. (1986), for the detail, see e.g. Haykin (1994)). However, it is now well-known that in practice the learning of the MLP-NN parameters by BP type algorithms quite often suffers from becoming stuck in a local minimum and requiring long period of learning in order to encode the training patterns, both of which are good reason for avoiding such networks in on-line processing.

This account also holds for training the ordinary radial basis function type networks (see e.g. Haykin, 1994) or self-organising feature maps (SOFMs) (Kohonen, 1997), since the network parameters tuning method resorts to a gradient-descent type algorithm, which normally requires iterative and long training (albeit some claims for the biological plausibility for SOFMs). A particular weakness of such networks is that when new training data arrives in on-line applications, an iterative learning algorithm must be reapplied to train the network from scratch using a combined the previous training and new data; i.e. incremental learning is generally quite hard.

2.2.2 Associative Memory/Hopfield's Recurrent Neural Networks

Associative memory has gained a great deal of interest for its structural resemblance to the cortical areas of the brain. In implementation, associative memory is quite often alternatively represented as a *correlation matrix*, since each neuron can be interpreted as an element of matrix. The data are stored in terms of a distributed representation, such as in MLP-NNs, and both the stimulus (key) and the response (the data) are required to form an associative memory.

In contrast, recurrent networks known as Hopfield-type recurrent neural networks (HRNNs) (Hopfield, 1982) are rooted in statistical physics and, as the name stands, have feedback connections. However, despite their capability to retrieve a stored pattern by giving only a reasonable subset of patterns, they also often suffer from becoming stuck in the so-called "spurious" states (Amit, 1989; Hertz et al., 1991; Haykin, 1994).

Both the associative memory and HRNNs have, from the mathematical view point, attracted great interest in terms of their dynamical behaviours. However, the actual implementation is quite often hindered in practice, due to the considerable amount of computation compared to feedforward artificial neural networks (Looney, 1997). Moreover, it is theoretically known that there is a storage limit, in which a Hopfield network cannot store more than $0.138N$ (N: total number of neurons in the network) random patterns, when it is used as a content-addressable memory (Haykin, 1994). In general, as for MLP-NNs, dynamic re-configuration of such networks is not possible, e.g. incremental learning when new data is arrived (Ritter et al., 1992).

In summary, conventional associative memory, HRNNs, MLP-NNs (see also Stork, 1989), RBF-NNs, and SOFMs are not that appealing as the candidates for modelling the learning mechanism of the brain (Roy, 2000).

2.2.3 Variants of RBF-NN Models

In relation to RBF-NNs, in disciplines other than artificial neural networks, a number of different models such as the generalised context model (GCM) (Nosofsky, 1986), the extended model called attention learning covering map (ALCOVE) (Kruschke, 1992) (both the GCM and ALCOVE were proposed in the psychological context), and Gaussian mixture model (GMM) (see e.g. Hastie et al., 2001) have been proposed by exploiting the property of a Gaussian response function. Interestingly, although these models all stemmed from disparate disciplines, the underlying concept is similar to that of the original RBF-NNs. Thus, within these models, the notion of weights between the nodes is still identical to RBF-NNs and rather arduous approximation of the weight parameters is thus involved.

2.3 PNNs and GRNNs

In the early 1990's, Specht rediscovered the effectiveness of kernel discriminant analysis (Hand, 1984) within the context of artificial neural networks. This led him to define the notion of a probabilistic neural network (PNN) (Specht, 1988, 1990). Subsequently, Nadaraya-Watson kernel regression (Nadaraya, 1964; Watson, 1964) was reformulated as a generalised regression neural network (GRNN) (Specht, 1991) (for a concise review of PNNs/GRNNs, see also

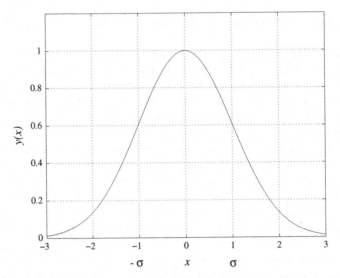

Fig. 2.1. A Gaussian response function: $y(x) = \exp(-x^2/2)$

(Sarle, 2001)). In the neural network context, both PNNs and GRNNs have layered structures as in MLP-NNs and can be categorised into a family of RBF-NNs (Wasserman, 1993; Orr, 1996) in which a hidden neuron is represented by a Gaussian response function.

Figure 2.1 shows a Gaussian response function:

$$y(x) = \exp\left(-\frac{x^2}{2\sigma^2}\right) \tag{2.1}$$

where $\sigma = 1$.

From the statistical point of view, the PNN/GRNN approach can also be regarded as a special case of a Parzen window (Parzen, 1962), as well as RBF-NNs (Duda et al., 2001).

In addition, regardless of minor exceptions, it is intuitively considered that the selection of a Gaussian response function is reasonable for the global description of the real-world data, as represented by the consequence from the *central limit theorem* in the statistical context (see e.g. Garcia, 1994).

Whilst the roots of PNNs and GRNNs differ from each other, in practice, the only difference between PNNs and GRNNs (in the strict sense) is confined to their implementation; for PNNs the weights between the RBFs and the output neuron(s) (which are identical to the target values for both PNNs and GRNNs) are normally fixed to binary (0/1) values, whereas GRNNs generally do not hold such restriction in the weight settings.

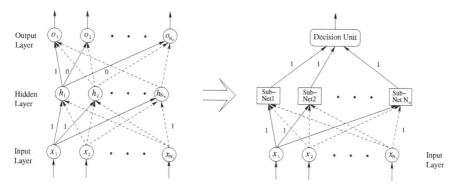

Fig. 2.2. Illustration of topological equivalence between the three-layered PNN/GRNN with N_h hidden and N_o output units and the assembly of the N_o distinct sub-networks

2.3.1 Network Configuration of PNNs/GRNNs

The left part in Fig. 2.2 shows a three-layered PNN (or GRNN with the binary weight coefficients between RBFs and output units) with N_i inputs, N_h RBFs, and N_o output units. In the figure, each input unit x_i ($i = 1, 2, \ldots, N_i$) corresponds to the element in the input vector $\mathbf{x} = [x_1, x_2, \ldots, x_{N_i}]^T$ (T: vector transpose), h_j ($j = 1, 2, \ldots, N_h$) is the j-th RBF (note that N_h is varied), $\|\ldots\|_2^2$ denotes the squared L_2 norm, and the output of each neuron o_k ($k = 1, 2, \ldots, N_o$) is calculated as[1]

$$o_k = \frac{1}{\xi} \sum_{j=1}^{N_h} w_{j,k} h_j \,, \tag{2.2}$$

where

$$\xi = \sum_{k=1}^{N_o} \sum_{j=1}^{N_h} w_{j,k} h_j \,,$$

$$\mathbf{w}_j = [w_{j,1}, w_{j,2}, \ldots, w_{j,N_o}]^T \,,$$

$$h_j = f(\mathbf{x}, \mathbf{c}_j, \sigma_j) = \exp\left(-\frac{\|\mathbf{x} - \mathbf{c}_j\|_2^2}{\sigma_j^2} \right) \,. \tag{2.3}$$

---

[1]In (2.2), the factor ξ is, in practice, used to *normalise* the resulting output values. Then, the manner given in (2.2) does not match the form derived originally from the conditionally probabilistic approach (Specht, 1990, 1991). However, in the original GRNN approach, the range of the output values depends upon the weight factor $w_{j,k}$ and is not always bounded within a certain range, which may not be convenient in the case of e.g. hardware representation. Therefore, the definition as in (2.2) is adopted in this book, since the *relative* values of the output neurons are given, instead of the original one.

In the above, c_j is called the centroid vector, σ_j is the radius, and $\mathbf{w}_j$ denotes the weight vector between the j-th RBF and the output neurons. In the case of a PNN, the weight vector $\mathbf{w}_j$ is given as a binary (0 or 1) sequence, which is identical to the target vector.

As in the left part of Fig. 2.2, the structure of a PNN/GRNN, at first examination, is similar to the well-known multilayered perceptron neural network (MLP-NN) except that RBFs are used in the hidden layer and linear functions in the output layer.

In comparison with the conventional RBF-NNs, the GRNNs have a special property, namely that *no iterative training of the weight vectors is required* (Wasserman, 1993). That is, as for other RBF-NNs, any input-output mapping is possible, by simply assigning the input vectors to the centroid vectors and fixing the weight vectors between the RBFs and outputs identical to the corresponding target vectors. This is quite attractive, since, as stated earlier, conventional MLP-NNs with back-propagation type weight adaptation involve long and iterative training, and there even may be a danger of becoming stuck in a local minimum (this is serious as the size of the training set becomes large).

Moreover, the special property of PNNs/GRNNs enables us to flexibly configure the network depending upon the tasks given, which is considered to be beneficial to real hardware implementation, with only two parameters, c_j and σ_j, to be adjusted. The only disadvantage of PNNs/GRNNs in comparison with MLP-NNs seems to be, due to the memory-based architecture, the need for storing all the centroid vectors into memory space, which can be sometimes excessive for on-line data processing, and hence, the operation is slow in the reference mode (i.e. the testing phase). Nevertheless, with the flexible configuration property, PNNs/GRNNs can be exploited for interpretation of the notions relevant to the actual brain.

In Fig. 2.2, when the target vector $\mathbf{t}(\mathbf{x})$ corresponding to the input pattern vector $\mathbf{x}$ is given as a vector of indicator functions

$$\mathbf{t}(\mathbf{x}) = (\delta_1, \delta_2, \ldots, \delta_{N_o}) ,$$

$$\delta_j = \begin{cases} 1 \; ; \; \text{if } \mathbf{x} \text{ belongs to the class} \\ \quad ; \; \text{corresponding to } o_k \\ 0 \; ; \; \text{otherwise.} \end{cases} \qquad (2.4)$$

and when the RBF h_j is assigned for $\mathbf{x}$, with utilising the special property of a PNN/GRNN, $\mathbf{w}_j = \mathbf{t}(\mathbf{x})$, the entire network becomes topologically equivalent to the network with a decision unit and N_o sub-networks as in the right part in Fig. 2.2.

In summary, the network configuration[2] by means of a PNN/GRNN is simply achieved as in the following:

---

[2]In the neural networks community, this configuration is often referred to as "learning". Strictly speaking, the usage of the terminology is, however, rather

---

[Summary of PNN/GRNN Network Configuration]

Network Growing :   Set $\mathbf{c}_j = \mathbf{x}$ and fix σ_j, then add the term $w_{jk}h_j$ in (2.3).

For pattern classification tasks, the target vector $\mathbf{t}(\mathbf{x})$ is thus used as a "class label", indicating the sub-network number to which the RBF belongs. (Namely, this operation is equivalent to add the j-th RBF in the corresponding (i.e. the k-th) Sub-Net in the left part in Fig. 2.2.)

Network Shrinking : Delete the term $w_{jk}h_j$ from (2.3).

In addition, by comparing a PNN with GRNN, it is considered that the weight setting of GRNNs may be exploited for a more flexible utility, e.g. in pattern classification problems, the fractional weight values can represent the "certainty" (i.e. the weights between the RBFs and output neurons are varied between zero to one, in accordance with the certainty of the RBF, by introducing a (sort of) fuzzy-logic decision scheme, by exploiting the *a priori* knowledge of the problem) that the RBF belongs to a particular class.

2.3.2 Example of PNN/GRNN – the Celebrated Exclusive OR Problem

As an example using a PNN/GRNN, let us consider the celebrated pattern classification problem of exclusive-or (XOR). This problem has quite often been treated as a benchmark for a pattern classifier, especially since Minsky and Papert (Minsky and Papert, 1969) proved the computational limitation of the simple Rosenblatt's perceptron model (Rosenblatt, 1958), which later led to the extension of the model to an MLP-NN; a perceptron cannot solve the XOR problem, since a perceptron essentially represents only a single separating line in the hyperplane, whilst for the solution to the XOR problem, (at least) two such lines are required.

Figure 2.3 shows the PNN/GRNN which gives a solution to the well-known exclusive-or (XOR) problem. In general, even to achieve the input-output relation of the simple XOR problem involves iterative tuning of the network node parameters by means of MLP-NNs, there is virtually no such iterative tuning involved in PNNs/GRNNs; in the case of an MLP-NN, two lines are needed to separate the circles filled with black (i.e. $y = 1$) from the other two ($y = 0$), as in Fig. 2.4 (a). In terms of an MLP-NN, it is equivalent that the properties of the two lines (i.e. both the slopes and y-intercepts) are tuned to provide such separation during the training. (Thus, it is evident that a single

---

limited, since the network is grown/shrunk by fixing the network parameters for a particular set of patterns other than *tuning* them, e.g. by repetitive adjustment of the weight vectors as in the ordinary back-propagation algorithm.

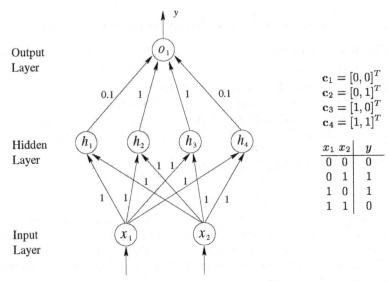

Fig. 2.3. A PNN/GRNN for the solution to the exclusive-or (XOR) problem – 1) the four units in the hidden layer (i.e. RBFs) h_i $(i = 1, 2, 3, 4)$ are assigned with fixing both the centroid vectors, $\mathbf{c}_1 = [0, 0]^T$, $\mathbf{c}_2 = [0, 1]^T$, $\mathbf{c}_3 = [1, 0]^T$, and $\mathbf{c}_4 = [1, 1]^T$, and (reasonably small values of) the radii and 2) the weights between the hidden and output layer are simply set to the four (values close to) target values, respectively, i.e. $w_{11} = 0.1$, $w_{12} = 1.0$, $w_{13} = 1.0$, and $w_{14} = 0.1$

perceptron cannot simultaneously provide two such separating lines.) On the other hand, as in Fig. 2.4 (b), when 1) the four hidden (or RBF) neurons h_i $(i = 1, 2, 3, 4)$ are assigned with fixing both the centroid vectors, $\mathbf{c}_1 = [0, 0]^T$, $\mathbf{c}_2 = [0, 1]^T$, $\mathbf{c}_3 = [1, 0]^T$, and $\mathbf{c}_4 = [1, 1]^T$, and (reasonably small values of) the radii and 2) the weights are simply set to the four (values close to) target values, respectively, i.e. $w_{11} = 0.1$, $w_{12} = 1.0$, $w_{13} = 1.0$, and, $w_{14} = 0.1$[3], the network tuning is completed (thus "one-pass" or "one-shot" training).

In the preliminary simulation study, the XOR problem was also solved by a three-layered perceptron NN; the network consists of only two nodes for both the input and hidden layers and one single output node. Then, the network was trained by the BP algorithm (Amari, 1967; Bryson and Ho, 1969; Werbos, 1974; Parker, 1985; Rumelhart et al., 1986) with a momentum term update scheme (Nakano et al., 1989) and tested using the same four patterns as aforementioned. However, as reported in (Nakano et al., 1989), it was empirically confirmed that the training of the MLP-NN requires (at least) some ten times of iterative weight adjustment, though the parameters were carefully chosen by trial and error, and thus that the "one-shot" training such

---

[3]Here, both the weight values $w_{11} = 0.1$ and $w_{14} = 0.1$ are considered, rather than $w_{11} = 0$ and $w_{14} = 0$, in order to keep the explicit network structure for the XOR problem.

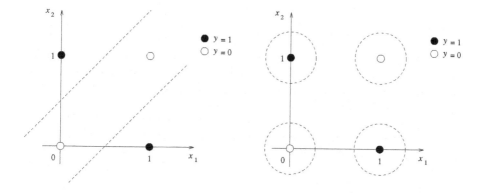

(a) Decision boundary repre-   (b) Decision boundary represented by an MLP-NN   sented by a PNN / GRNN

Fig. 2.4. Comparison of decision boundaries for (**a**) an MLP-NN and (**b**) PNN/GRNN for the solution to the XOR – in the case of an MLP-NN, two lines are needed to separate the circles (i.e. $y = 1$ filled with black) from the other two ($y = 0$), whilst the decision boundaries for a PNN/GRNN are determined by the four RBFs

as PNNs/GRNNs can never be achieved using the MLP-NN, even for this small task.

2.3.3 Capability in Accommodating New Classes within PNNs/GRNNs (Hoya, 2003a)

In Hoya (2003a), it is reported that a PNN exhibits a capability to accommodate new classes, whilst maintaining a reasonably high generalisation capability. In essence, this feature is particularly important and desirable for pattern classification tasks.

In a recent study (Polikar et al., 2001), a new guideline for the incremental learning paradigm in pattern classification has been given in accordance with the four criteria:

    1) The pattern classifier(s) should be able to learn additional information from the new data;

    2) They should not require access to the original data used to train the existing classifier;

    3) They should preserve previously acquired knowledge (that is, they should not suffer from catastrophic forgetting);

    4) They should be able to accommodate new classes that may be introduced within the new data.

It is then obvious that the network growing phase within the network configuration rule given earlier suffices the criterion 1) above, since a newly incoming pattern vector can be readily assigned to the centroid vector of a new RBF and thereby since a new local pattern space is formed within the entire space already given. Then, it is intuitively said that Criterion 3) above can also be satisfied, unless the local pattern space so formed does not seriously pervade (but may moderately overlap) other local spaces.

Thus, from the structural point of view, accommodating new classes is nothing more than simply adding a cluster of RBF(s) or, in other words, new subnets within the PNN/GRNN. However, this is possible, under the assumption that one pattern space spanned by a subnet is *reasonably* separated from the others.

Accordingly, Criterion 4) above can be satisfied in terms of PNNs/GRNNs, which will be justified in the simulation examples given later.

2.3.4 Necessity of Re-accessing the Stored Data

Up to now, what remains is Criterion 2), regarding the requirement of accessing the original data to train the existing classifier.

In Polikar et al. (2001), the authors pointed out that supervised networks such as adaptive resonance theory maps (ARTMAPs) (Carpenter, 1991) suffer from poor generalisation capability due to over-fitting, at the expense of no access to the previously seen data. To overcome this drawback, it is generally necessary to involve either the *a priori* knowledge (e.g. data distributions) or a sort of *ad hoc* parameter adjustment scheme. A similar principle also applies to the case of a PNN; in order to maintain the good generalisation capability, the internal access to the stored data is necessary so as to update the radii values. However, one of the key advantages using the PNN is that, since a PNN represents a memory-based architecture, it does not require storage of entire original data besides the memory space for the PNN itself. In other words, (some of) the original data are directly accessible via the internally stored data, i.e. the centroid vectors c_j. In practice, Criterion 2) above is therefore too strict and hence re-accessing the original data is still unavoidable. However, as described later in this book, this could also be circumvented in terms of the modular architecture (albeit different from conventional modular neural networks) approach, within the kernel memory principle.

2.3.5 Simulation Example

In Hoya (2003a), a simulation example using four benchmark datasets for pattern classification is given to show the capability in accommodating new classes within a PNN; the speech filing system (SFS) (Huckvale, 1996) for digit voice classification (i.e. /ZERO/, /ONE/, ..., /NINE/, in English) and the three UCI data sets, which are chosen from "UCI Machine Learning

Repository" of the University of California[4], namely the OptDigit, PenDigit, and ISOLET data set are employed. For the SFS data set, each utterance was firstly encoded by the commonly used linear predictive coding (LPC) mel-cepstral analysis(see e.g. Deller et al., 1993; Furui, 1981) for speech coding[5] and given as a feature vector with 256 data points. For the three UCI data sets, the first two are used for digit character recognition tasks, whilst the latter is for isolated letter speech recognition tasks, all of which are ready for performing the pattern classification. The description of the data sets used is summarised in Table 2.1.

Table 2.1. Data sets used to show the capability in accommodating new classes within a PNN

Data Set	Length of Each Pattern Vector	Total Num. of Patterns in the Training Set	Total Num. of Patterns in the Testing Sets	Num. of Classes
SFS	256	599	400	10
OptDigit	64	3823	1797	10
PenDigit	16	7494	3498	10
ISOLET	617	1040	520	26

Performance Measurement

To investigate the capability of accommodating new classes within a PNN, the measurement in terms of deterioration rate d, which is given as the difference in the number of correctly classified patterns between the initially configured PNN and its grown version with new classes, is introduced as

$$d = \frac{c_i - c_g}{N} \tag{2.5}$$

[4]The original datasets, OptDigit, PenDigit, and ISOLET, were downloaded from UCI Machine Learning Repository at: http://www.ics.uci.edu/mlearn/MLRepository.html

[5]More specifically, the original utterances in the SFS dataset were sampled at 20 kHz and each utterance was firstly pre-processed using a seventh-order adaptive inverse filter (Nakajima et al., 1978). Second, the entire sample sequence was converted into 16 uniformly allocated frames (overlapping or distinct, depending upon both the lengths of an analysis window frame and the whole sequence). Then, each frame data was transformed into the power-spectrum domain by applying the LPC mel-cepstral analysis (Furui, 1981) with 14 coefficients. The power-spectrum domain data (or the power spectral density, PSD) points (per frame) were further converted into 16 data points by smoothing the power-spectrum (i.e. applying a low-pass filter operation). Finally, for each utterance, a total of 256(= 16 frames × 16 points) data points were obtained and used as the feature vector of the pattern classifier.

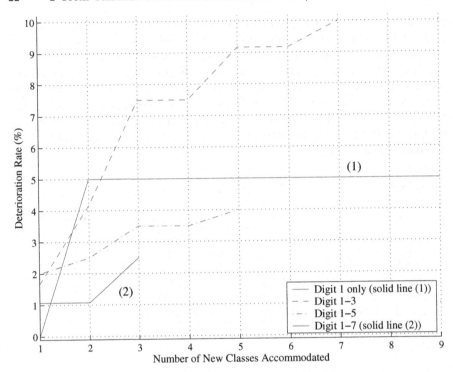

Fig. 2.5. Transition of the deterioration rate with varying the number of new classes accommodated – SFS data set

where

  c_i: number of correctly classified patterns with the initial configuration;
  c_g: number of correctly classified patterns with the grown network;
  N: total number of testing patterns.

Note that, for the computation of (2.5), to give a fair comparison, the total number of testing patterns N was also varied according to the number of *initially* accommodated classes (digits/letters).

Simulation Results

In Figs. 2.5–2.8, each of the four lines shows the transition of the deterioration rate (defined in (2.5)) obtained by varying the number of new classes (digits/letters) accommodated within the original PNN. In each figure, the label "Digit i-j" (or "Letter i-j" for ISOLET) indicates that the PNN was initially configured with the pattern vectors for *only* the classes from Digit/Letter i to j. For all the data sets, the overall generalisation performance (using the testing set) with the initial configuration remained satisfactory, i.e. within the range from 90.4% to 100.0%.

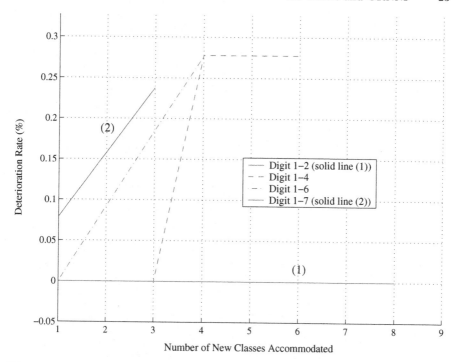

Fig. 2.6. Transition of the deterioration rate with varying the number of new classes accommodated – OptDigit data set

Discussion

For all the cases, a similar tendency was observed; as the number of new classes is increased, the generalisation performance deteriorates. This naturally follows, since the number of degrees of freedom is also increased by adding new classes to be classified. However, as in Figs. 2.5–2.8, this is also dependent upon the length of the pattern vectors and was confirmed by another set of simulations; with identical numbers of the patterns in both the training and testing sets (i.e. 200 for training and 300 for testing) for the three data sets, the SFS, OptDigit, and PenDigit (of which the number of classes is also identical), the overall deterioration rate of PenDigit was less than that of the SFS. In other words, this observation indicates that the coverage of pattern space by the RBFs is accordingly broadened as the dimensionality is decreased.

In addition, for the PenDigit case (i.e. using the original large data set of PenDigit), a deterioration rate of around 14% was observed for "Digit 1–2", by increasing the number of classes to three. In such a case, it can be said that "over-training" may have occurred, due to the excessive amount of the training data, by taking into account the length of each pattern vector (i.e. 16). This indicates that, as for other neural based pattern classifiers, pruning

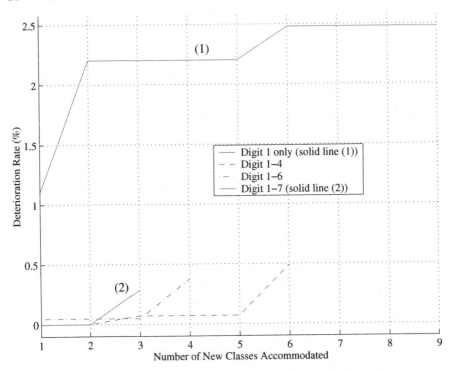

Fig. 2.7. Transition of the deterioration rate with varying the number of new classes accommodated – PenDigit data set

of the training data in advance is important for the training (or constructing) of a PNN (for a further discussion of this, see e.g. Hoya, 1998).

Then, as shown (*solid lines*) in Figs. 2.5–2.8, the deterioration rate of the initial configuration with the smaller number of classes (i.e. trained only with either one or two classes) was, as expected, the highest for the three data sets, i.e. SFS, PenDigit, and ISOLET. This can be interpreted such that the separation of the pattern space with a smaller number of classes is rather broad and thus is easily eroded by adding new classes. This erosion was noticeable in the case of ISOLET. However, it can also be said that the degree of erosion is more or less bounded. In other words, the spread of the RBFs is limited, since, as shown in Figs. 2.5–2.8, the deterioration rate remained the same when the number of classes was increased. In this regard, it is considered that the structure of the training data set for OptDigit is most well-balanced amongst the four, since the deterioration rate was low (which was less than 0.3%), whereas the generalisation performance was relatively high, i.e. around 99.0% for all the initial conditions. In contrast, for SFS, the deterioration rate was rather steadily increased as the number of new classes for all the initial configurations (except the case "Digit 1 only") was increased, in comparison

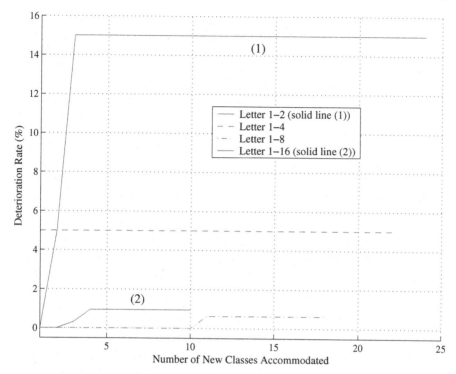

Fig. 2.8. Transition of the deterioration rate with varying the number of new classes accommodated – ISOLET data set

with the other three data sets. This is perhaps due to the insufficient number of pattern vectors and thereby the weak coverage of the pattern space.

Nevertheless, it is stated that, by exploiting the flexible configuration property of a PNN, the separation of pattern space can be kept sufficiently well for each class even when adding new classes, as long as the amount of the training data is not excessive for each class. Then, as discussed above, this is supported by the empirical fact that the generalisation performance was not seriously deteriorated for almost all the cases.

It can therefore be concluded that any "catastrophic" forgetting of the previously stored data due to accommodation of new classes did not occur, which meets Criterion 4).

2.4 Comparison Between Commonly Used Connectionist Models and PNNs/GRNNs

In practice, the advantage of PNNs/GRNNs is that they are essentially free from the "baby-sitting" required for e.g. MLP-NNs or SOFMs, i.e. the necessity to tune a number of network parameters to obtain a good convergence rate or worry about any numerical instability such as local minima or long and iterative training of the network parameters. As described earlier, by exploiting the property of PNNs/GRNNs, simple and quick incremental learning is possible due to their inherently memory-based architecture[6], whereby the network growing/shrinking is straightforwardly performed (Hoya and Chambers, 2001a; Hoya, 2004b).

In terms of the generalisation capability within the pattern classification context, PNNs/GRNNs normally exhibit similar capability as compared with MLP-NNs; in Hoya (1998), such a comparison using the SFS dataset is made, and it is reported that a PNN/GRNN with the same number of hidden neurons as an MLP-NN yields almost identical classification performance. Related to this observation, in Mak et al. (1994), Mak et al. also compared the classification accuracy of an RBF-NN with an MLP-NN in terms of speaker identification and concluded that an RBF-NN with appropriate parameter settings could even surpass the classification performance obtained by an MLP-NN.

Moreover, as described, by virtue of the flexible network configuration property, adding new classes can be straightforwardly performed, under the assumption that one pattern space spanned by a subnet is *reasonably* separated from the others. This principle is particularly applicable to PNNs and GRNNs; the training data for other widely-used layered networks such as MLP-NNs trained by a back-propagation algorithm (BP) or ordinary RBF-NNs is encoded and stored within the network after the iterative learning. On the other hand, in MLP-NNs, the encoded data are then distributed over the weight vectors (i.e. *sparse* representation of the data) between the input and hidden layers and those between hidden and output layers (and hence not directly accessible).

Therefore, it is generally considered that, not to mention the accommodation of new classes, to achieve a flexible network configuration by an MLP-NN similar to that by a PNN/GRNN (that is, the quick network growing and shrinking) is very hard. This is because even a small adjustment of the weight parameters will cause a dramatic change in the pattern space constructed, which may eventually lead to a catastrophic corruption of the pattern space (Polikar et al., 2001). For the network reconfiguration of MLP-NNs, it is thus normally necessary for the iterative training to start from scratch. From another point of view, by MLP-NNs, the separation of the pattern space is represented in terms of the hyperplanes so formed, whilst that performed by PNNs and GRNNs is based upon the location and spread of the RBFs in the pattern space. In PNNs/GRNNs, it is therefore considered that, since a single class is essentially represented by a cluster of RBFs, a small change in a particular cluster does not have any serious impact upon other classes, unless the spread of the RBFs pervades the neighbour clusters.

---

[6]In general, the original RBF-NN scheme has already exhibited a similar property; in Poggio and Edelman (1990), it is stated that a reasonable initial performance can be obtained by merely setting the centres (i.e. the centroid vectors) to a subset of the examples.

Table 2.2. Comparison of symbol-grounding approaches and feedforward type networks – GRNNs, MLP-NNs, PNNs, and RBF-NNs

	Symbol Processing Approaches	Generalised Regression Neural Networks **(GRNN)/** Probabilistic Neural Networks **(PNN)**	Multilayered Perceptron Neural Networks (MLP-NN)/Radial Basic Function Neural Networks (RBF-NN)
Data Representation	Not Encoded	**Not Encoded**	Encoded
Straightforward Network Growing/ Shrinking	Yes	**Yes**	No (Yes for RBF-NN)
Numerical Instability	No	**No**	Yes
Memory Space Required	Huge	**Relatively Large**	Moderately Large
Capability in Accommodating New Classes	Yes	**Yes**	No

In Table 2.2, a comparison of commonly used layered type artificial neural networks and symbol-based connectionist models is given, i.e. symbol processing approaches as in traditional artificial intelligence (see e.g. Newell and Simon, 1997) (where each node simply consists of the pattern and symbol (label) and no further processing between the respective nodes is involved) and layered type artificial neural networks, i.e. GRNNs, MLP-NNs, PNNs, and RBF-NNs.

As in Table 2.2 and the study (Hoya, 2003a), the disadvantageous points of PNNs may, in turn, reside in 1) the necessity for relatively large space in storing the network parameters, i.e. the centroid vectors, 2) intensive access to the stored data within the PNNs in the reference (i.e. testing) mode, 3) determination of the radii parameters, which is relevant to 2), and 4) how to determine the size of the PNN (i.e. the number of hidden nodes to be used).

In respect of 1), MLP-NNs seem to have an advantage in that the distributed (or sparse) data representation obtained after the learning may yield a more compact memory space than that required for PNN/GRNN, albeit at the expense of iterative learning and the possibility of the aforementioned numerical problems, which can be serious, especially when the size of the training set is large. However, this does not seem to give any further advantage, since, as in the pattern classification application (Hoya, 1998), an RBF-NN (GRNN) with the same size of MLP-NN may yield a similar performance.

For 3), although some iterative tuning methods have been proposed and investigated (see e.g. Bishop, 1996; Wasserman, 1993), in Hoya and Chambers

(2001a); Hoya (2003a, 2004b), it is reported that a unique setting of the radii for all the RBFs, which can also be regarded as the modified version suggested in (Haykin, 1994), still yields a reasonable performance:

$$\sigma_j = \sigma = \theta_\sigma \times d_{max} \,, \tag{2.6}$$

where d_{max} is maximum Euclidean distance between all the centroid vectors within a PNN/GRNN, i.e. $d_{max} = \max(\|\mathbf{c}_l - \mathbf{c}_m\|_2^2), (l \neq m)$, and θ_σ is a suitably chosen constant (for all the simulation results given in Sect. 2.3.5, the setting $\theta_\sigma = 0.1$ was employed.) Therefore, this is not considered to be crucial.

Point 4) still remains an open issue related to pruning of the data points to be stored within the network (Wasserman, 1993). However, the selection of data points, i.e. the determination of the network size, is not an issue limited to the GRNNs and PNNs. MacQueen's k-means method (MacQueen, 1967) or, alternatively, graph theoretic data-pruning methods (Hoya, 1998) could be potentially used for clustering in a number of practical situations. These methods have been found to provide reasonable generalisation performance (Hoya and Chambers, 2001a). Alternatively, this can be achieved by means of an intelligent approach, i.e. within the context of the evolutionary process of a hierarchically arranged GRNN (HA-GRNN) (to be described in Chap. 10), since, as in Hoya (2004b), the performance of the sufficiently evolved HA-GRNN is superior to an ordinary GRNN with exactly the same size using MacQueen's k-means clustering method. (The issues related to HA-GRNNs will be given in more detail later in this book.)

Thus, the most outstanding issue pertaining to a PNN/GRNN seems to be 2). However, as described later (in Chap. 4), in the context of the self-organising kernel memory concept, this may not be such an issue, since, during the training phase, just one-pass presentation of the input data is sufficient to self-organise the network structure. In addition, by means of the modular architecture (to be discussed in Chap. 8; the hierarchically layered long-term memory (LTM) networks concept), the problem of intensive access, i.e. to update the radii values, could also be solved.

In addition, with a supportive argument regarding the RBF units in Vetter et al. (1995), the approach in terms of RBFs (or, in a more general term, the kernels) can also be biologically appealing. It is then fair to say that the functionality of an RBF unit somewhat represents that of the so-called "grand-mother' cells (Gross et al., 1972; Perrett et al., 1982)[7]. (We will return to this issue in Chap. 4.)

---

[7]However, at the neuro-anatomical level, whether or not such cells actually exist in a real brain is still an open issue and beyond the scope of this book. Here, the author simply intends to highlight the importance of the neurophysiological evidence that some cells (or the column structures) may represent the *functionality* of the "grandmother" cells which exhibit such generalisation capability.

2.5 Chapter Summary

In this chapter, a number of artificial neural network models that stemmed from various disciplines of connectionism have firstly been reviewed. It has then been described that the three inherent properties of the PNNs/GRNNs:

- Straightforward network (re-)configuration (i.e. both network growing and shrinking) and thus the utility in time-varying situations;
- Capability in accommodating new classes (categories);
- Robust classification performance which can be comparable to/exceed that of MLP-NNs (Mak et al., 1994; Hoya, 1998)

are quite useful for general pattern classification tasks. These properties have been justified with extensive simulation examples and compared with commonly-used connectionist models.

The attractive properties of PNNs/GRNNs have given a basis for modeling psychological functions (Hoya, 2004b), in which the psychological notion of memory dichotomy (James, 1890) (to be described later in Chap. 8), i.e. the neuropsychological speculation that conceptually the memory should be divided into short- and long-term memory, depending upon the latency, is exploited for the evolution of a hierarchically arranged generalised regression neural network (HA-GRNN) consisting of a multiple of modified generalised regression neural networks and the associated learning mechanisms (in Chap. 10), namely a framework for the development of brain-like computers (cf. Matsumoto et al., 1995) or, in a more realistic sense of, "artificial intelligence". The model and the dynamical behaviour of an HA-GRNN will be more informatively described later in this book.

In summary, on the basis of the remarks in Matsumoto et al. (1995), it is considered that the aforementioned features of PNNs/GRNNs are fundamentals to the development of brain-like computers.

2.8 Chapter Summary

3
_____

The Kernel Memory Concept – A Paradigm Shift from Conventional Connectionism

3.1 Perspective

In this chapter, the general concept of kernel memory (KM) is described, which is given as the basis for not only representing the general notion of "memory" but also modelling the psychological functions related to the artificial mind system developed in later chapters.

As discussed in the previous chapter, one of the fundamental reasons for the numerical instability problem within most of conventional artificial neural networks lies in the fact that the data are encoded within the weights between the network nodes. This particularly hinders the application to on-line data processing, as is inevitable for developing more realistic brain-like information systems.

In the KM concept, as in the conventional connectionist models, the network structure is based upon the network nodes (i.e. called the _kernels_) and their connections. For representing such nodes, any function that yields the output value can be applied and defined as the _kernel function_. In a situation, each kernel is defined and functions as a similarity measurement between the data given to the kernel and memory stored within. Then, unlike conventional neural network architectures, the "weight" (alternatively called _link weight_) between a pair of nodes is redefined to simply represent the _strength_ of the connection between the nodes. This concept was originally motivated from a neuropsychological perspective by Hebb (Hebb, 1949), and, since the actual data are encoded _not_ within the weight parameter space but within the _template_ vectors of the kernel functions (KFs), the tuning of the weight parameters does not dramatically affect the performance.

3.2 The Kernel Memory

In the kernel memory context, the most elementary unit is called a single _kernel unit_ that represents the local memory space. The term _kernel_ denotes

Tetsuya Hoya: _Artificial Mind System – Kernel Memory Approach_, Studies in Computational Intelligence (SCI) **1**, 31–58 (2005)
www.springerlink.com

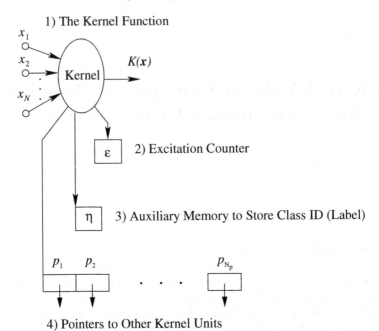

Fig. 3.1. The kernel unit – consisting of four elements; given the inputs $\mathbf{x} = [x_1, x_2, \ldots, x_N]$ 1) the kernel function $K(\mathbf{x})$, 2) an excitation counter ε, 3) auxiliary memory to store the class ID (label) η, and 4) pointers to other kernel units p_i $(i = 1, 2, \ldots, N_p)$

a *kernel function*, the name of which originates from integral operator theory (see Christianini and Taylor, 2000). Then, the term is used in a similar context within kernel discriminant analysis (Hand, 1984) or kernel density estimation (Rosenblatt, 1956; Jutten, 1997), also known as Parzen windows (Parzen, 1962), to describe a certain distance metric between a pair of vectors. Recently, the name *kernel* has frequently appeared in the literature, essentially on the same basis, especially in the literature relevant to support vector machines (SVMs) (Vapnik, 1995; Hearst, 1998; Christianini and Taylor, 2000).

Hereafter in this book, the terminology *kernel* [1] is then frequently referred to as (but not limited to) the kernel function $K(\mathbf{a}, \mathbf{b})$ which merely represents a certain distance metric between two vectors $\mathbf{a}$ and $\mathbf{b}$.

3.2.1 Definition of the Kernel Unit

Figure 3.1 depicts the kernel unit used in the kernel memory concept. As in the figure, a single kernel unit is composed of 1) the kernel function, 2)

---

[1]In this book, the term *kernel* sometimes interchangeably represents "kernel unit".

excitation counter, 3) auxiliary memory to store the class ID (label), and 4) pointers to the other kernel units.

In the figure, the first element, i.e. the *kernel function* $K(\mathbf{x})$ is formally defined:

$$K(\mathbf{x}) = f(\mathbf{x}) = f(x_1, x_2, \ldots, x_N) \tag{3.1}$$

where $f(\cdot)$ is a certain function, or, if it is used as a similarity measurement in a specific situation:

$$K(\mathbf{x}) = K(\mathbf{x}, \mathbf{t}) = D(\mathbf{x}, \mathbf{t}) \tag{3.2}$$

where $\mathbf{x} = [x_1, x_2, \ldots, x_N]^T$ is the input vector to the new memory element (i.e. a kernel unit), $\mathbf{t}$ is the template vector of the kernel unit, with the same dimension as $\mathbf{x}$ (i.e. $\mathbf{t} = [t_1, t_2, \ldots, t_N]^T$), and the function $D(\cdot)$ gives a certain metric between the vector $\mathbf{x}$ and $\mathbf{t}$.

Then, a number of such kernels as defined by (3.2) can be considered. The simplest of which is the form that utilises the Euclidean distance metric:

$$K(\mathbf{x}, \mathbf{t}) = \|\mathbf{x} - \mathbf{t}\|_2^n \ (n > 0) \ , \tag{3.3}$$

or, alternatively, we could exploit a variant of the basic form (3.3) as in the following table (see e.g. Hastie et al., 2001):

Table 3.1. Some of the commonly used kernel functions

Inner product:

$$K(\mathbf{x}) = K(\mathbf{x}, \mathbf{t}) = \mathbf{x} \cdot \mathbf{t} \tag{3.4}$$

Gaussian:

$$K(\mathbf{x}) = K(\mathbf{x}, \mathbf{t}) = \exp\left(-\frac{\|\mathbf{x} - \mathbf{t}\|^2}{\sigma^2}\right) \tag{3.5}$$

Epanechnikov quadratic:

$$K(\mathbf{x}) = K(z) = \begin{cases} \frac{3}{4}(1 - z^2) & \text{if } |z| < 1; \\ 0 & \text{otherwise} \end{cases} \tag{3.6}$$

Tri-cube:

$$K(\mathbf{x}) = K(z) = \begin{cases} (1 - |z|^3)^3 & \text{if } |z| < 1; \\ 0 & \text{otherwise} \end{cases} \tag{3.7}$$

where $z = \|\mathbf{x} - \mathbf{t}\|^n \ (n > 0)$.

The Gaussian Kernel

In (3.2), if a Gaussian response function is chosen for a kernel unit, the output of the kernel function $K(\mathbf{x})$ is given as[2]

$$K(\mathbf{x}) = K(\mathbf{x}, \mathbf{c}) = \exp\left(-\frac{\|\mathbf{x} - \mathbf{c}\|^2}{\sigma^2}\right). \qquad (3.8)$$

In the above, the template vector $\mathbf{t}$ is replaced by the centroid vector $\mathbf{c}$ which is specific to a Gaussian response function.

Then, the kernel function represented in terms of the Gaussian response function exhibits the following properties:

1) The distance metric between the two vectors $\mathbf{x}$ and $\mathbf{c}$ is given as the squared value of the Euclidean distance (i.e. the L_2 norm).
2) The *spread* of the output value (or, the width of the kernel) is determined by the factor (radius) σ.
3) The output value obtained by calculating $K(\mathbf{x})$ is *strictly* bounded within the range from 0 to 1.
4) In terms of the Taylor series expansion, the exponential part within the Gaussian response function can be approximated by the polynomial

$$\exp(-z) \approx \sum_{n=0}^{N} \frac{(-1)^n z^n}{n!}$$

$$= 1 - z + \frac{1}{2}z^2 - \frac{1}{3!}z^3 + \cdots \qquad (3.9)$$

where N is finite and reasonably large in practice. Exploiting this may facilitate hardware representation[3]. Along with this line, it is reported in (Platt, 1991) that the following approximation is empirically found to be reasonable:

$$\exp(-\frac{z}{\sigma^2}) \approx \begin{cases} (1 - (\frac{z}{q\sigma^2})^2)^2 & \text{if } z < q\sigma^2; \\ 0 & \text{otherwise} \end{cases} \qquad (3.10)$$

where $q = 2.67$.
5) The real world data can be moderately but reasonably well-represented in many situations in terms of the Gaussian response function, i.e. as a consequence of the central limit theorem in the statistical sense (see e.g.

---

[2]In some literature, the factor σ^2 within the denominator of the exponential function in (3.8) is multiplied by 2, due to the derivation of the original form. However, there is essentially no difference in practice, since we may rewrite (3.8) with $\sigma = \sqrt{2}\acute{\sigma}$, where $\acute{\sigma}$ is then regarded as the radius.

[3]For the realisation of the Gaussian response function (or RBF) in terms of hardware, the complimentary metal-oxide semiconductor (CMOS) inverters have been exploited (for the detail, see Anderson et al., 1993; Theogarajan and Akers, 1996, 1997; Yamasaki and Shibata, 2003).

Garcia, 1994) (as described in Sect. 2.3). Nevertheless, within the kernel memory context, it is also possible to use *a mixture of kernel represen-tations* rather than resorting to a single representation, depending upon situations.

In 1) above, a single Gaussian kernel is already a pattern classifier in the sense that calculating the Euclidean distance between **x** and **c** is equiva-lent to performing pattern matching and then the score indicating how *similar* the input vector **x** is to the stored pattern **c** is given as the value obtained from the exponential function (according to 3) above); if the value becomes asymptotically close to 1 (or, if the value is above a certain threshold), this indicates that the input vector **x** given matches the template vector **c** to a great extent and can be classified as the same category as that of **c**. Otherwise, the pattern **x** belongs to another category[4].

Thus, since the value obtained from the similarity measurement in (3.8) is bounded (or, in other words, normalised), due to the existence of the exponen-tial function, the uniformity in terms of the classification score is retained. In practice, this property is quite useful, especially when considering the utility of a multiple of Gaussian kernels, as used in the family of RBF-NNs. In this context, the Gaussian metric is advantageous in comparison with the original Euclidean metric given by (3.3).

Kernel Function Representing a General Symbolic Node

In addition, a single kernel can also be regarded as a new entity in place of the conventional memory element, as well as a symbolic node in general symbolism by simply assigning the kernel function as

$$K(\mathbf{x}) = \begin{cases} \theta_s & ; \text{if the activation from the other kernel} \\ & \text{unit(s) is transferred to this kernel} \\ & \text{unit via the link weight(s)} \\ 0 & ; \text{otherwise} \end{cases} \tag{3.11}$$

where θ_s is a certain constant.

This view then allows us to subsume the concept of symbolic connectionist models such as Minsky's knowledge-line (K-Line) (Minsky, 1985). Moreover, the kernel memory can replace the ordinary symbolism in that each node (i.e. represented by a single kernel unit) can have a generalisation capability which could, to a greater extent, mitigate the "curse-of-dimensionality", in which, practically speaking, the exponentially growing number of data points soon exhausts the entire memory space.

---

[4]In fact, the utility of Gaussian distribution function as a similarity measurement between two vectors is one of the common techniques, e.g. the psychological model of GCM (Nosofsky, 1986), which can be viewed as one of the twins of RBF-NNs, or the application to continuous speech recognition (Lee et al., 1990; Rabiner and Juang, 1993).

The Excitation Counter

Returning to Fig. 3.1, the second element of the kernel unit ε is the *excitation counter*. The excitation counter can be used to count how many times the kernel unit is *repeatedly excited* (e.g. the value of the kernel function $K(\mathbf{x})$ is above a given threshold) in a certain period of time (if so defined), i.e. when the kernel function satisfies the relation

$$K(\mathbf{x}) \geq \theta_K \tag{3.12}$$

where θ_K is the given threshold.

Initially, the value ε is set to 0 and incremented whenever the kernel unit is excited, though the value may be reset to 0, where necessary.

The Auxiliary Memory

The third element in Fig. 3.1 is the *auxiliary memory* η to store the class ID (label) indicating that the kernel unit belongs to a particular class (or category). Unlike the conventional pattern classification context, the timing to fix the class ID η is flexibly determined, which is dependent upon the learning algorithm for the kernel memory, as described later.

The Pointers to Other Kernel Units

Finally, the fourth element in Fig. 3.1 is the pointer(s) p_i $(i = 1, 2, \ldots, N_p)$ to the other kernel unit(s). Then, by exploiting these pointers, the *link weight*, which lies between a pair of kernel units with a weighting factor to represent the *strength of the connection* in between, is given.

> Note that this manner of connection then allows us to realise a different form of network configuration from the conventional neural network architectures, since the output of the kernel function $K(\mathbf{x})$ is not always directly transferred to the other nodes via the "weights", e.g. those between the hidden and output layers, as in PNNs/GRNNs.

3.2.2 An Alternative Representation of a Kernel Unit

It is also possible to design the kernel memory in such a way that, instead of introducing the class label η attached to each kernel unit, the kernel units are connected to the unit(s) which represents a class label (as the output node in PNNs/GRNNs or conventional symbolic networks), whilst keeping the same functionality as a memory element. (Then, this also implies that the kernel units representing class IDs/labels can be formed, or dynamically varied, during the course of the learning, as described in Chap. 7.) In such a case, the

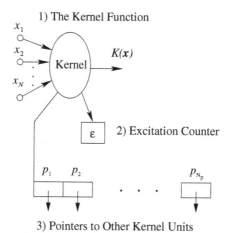

Fig. 3.2. A representation of a kernel unit (without the auxiliary memory) alternative to that in Fig. 3.1

kernel unit representation depicted in Fig. 3.2 can be more appropriate, instead of that in Fig. 3.1. (In the figure, note that the auxiliary memory for η is removed.)

Then, a single kernel unit is allowed to belong to multiple classes/categories at a time (Greenfield, 1995) by having kernels indicating the respective categories (or classes) and exploiting the pointers to other kernels in order to make connections in between. For instance, this allows a kernel unit to represent the word "penguin" both classified as English and Japanese.

In addition, the alternative kernel representation as shown in Fig. 3.2 can be more flexible; provided that a kernel representing a class ID is given and that the class ID is varied from the original, it is sufficient to change only the parameters of the kernel representing the class ID, which can in practice be more efficient. Thus, for this case, there is no need to alter the content of the auxiliary memory η for all the kernels that belong to the same class. (Then, the extension to the case of multiple class IDs is straightforward.)

3.2.3 Reformation of a PNN/GRNN

By exploiting the three elements within the kernel unit as illustrated in Fig. 3.1, i.e. 1) the kernel function, 2) the auxiliary memory to store the class ID of the kernel, and 3) the pointers to other kernel units, the PNNs/GRNNs can be reformulated as special cases of kernel memory with the three constraints on the network structure, namely, 1) only a single layer of Gaussian kernels is used but no lateral connections in a layer are allowed, 2) another layer for giving the results is provided, and 3) the two (i.e. the hidden and output) layers are fully-connected (allowing fractional weight values in the case of

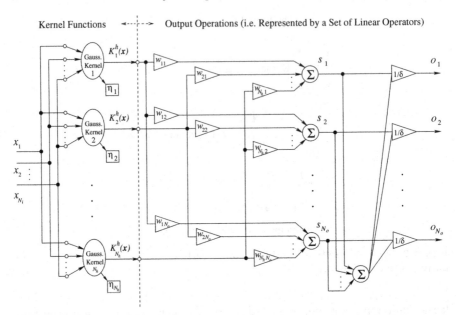

Fig. 3.3. A PNN/GRNN represented in terms of a set of the Gaussian kernels K_i^h (h: "hidden" layer, $i = 1, 2, \ldots, N_h$, with the auxiliary memory η_i to store the class ID but devoid of both the excitation counters and pointers to the other kernel units) and linear operators eventually yielding the outputs

GRNNs). In this context, a three-layered PNN/GRNN, which has previously been defined in the form (2.2) and (2.3) in Sect. 2.3.1, is equivalent to the kernel memory structure with multiple Gaussian kernels and the kernels with linear operations, the latter of which represent the respective output units.

First of all, as depicted in Fig. 3.3, a PNN/GRNN can be divided into two parts within the kernel memory concept; 1) a collection of Gaussian kernel units K_i^h (h: the kernels in the "hidden" layer, $i = 1, 2, \ldots, N_h$, with the auxiliary memory η_i but devoid of both the excitation counters and pointers to the other kernel units, e.g. for the lateral connections) and 2) (post-)linear output operations. Then, the former converts the input space into another domain in terms of the Gaussian kernel functions and the conversion is nonlinear, whilst the latter is based upon the linear operations in terms of both scaling and summation.

In Fig. 3.3, the scaling factor (or link weight) w_{ij} between the i-th Gaussian kernel and the j-th summation operator s_j with the activation $K_i^h(\mathbf{x})$

$$K_i^h(\mathbf{x}) = \exp\left(-\frac{\|\mathbf{x} - \mathbf{c}_i\|^2}{\sigma_i^2}\right), \tag{3.13}$$

where $\mathbf{x} = [x_1, x_2, \ldots, x_{N_i}]^T$, is identical to the corresponding element of the target vector, as described in Sect. 2.3.1. Then, the output value from the neuron o_j is given as a normalised linear sum:

$$s_j = \sum_{i=1}^{N_h} w_{ij} K_i^h(\mathbf{x})$$

$$o_j = f_1(\mathbf{x}) = \frac{1}{\xi} s_j \tag{3.14}$$

where ξ is a constant for normalising the output values and may be given as that in (2.3).

In PNNs/GRNNs, however, since it is evident that all the Gaussian kernels are eventually connected to the linear sum operators without any other lateral connections, the third elements, i.e. the pointers to other kernels, are omitted from the figure[5].

In general, if a multiple of output neurons o_j $(j = 1, 2, \ldots, N_o)$ are defined for pattern classification tasks, the final result will be obtained by choosing the output neuron with a maximum activation, which is the so-called "winner-takes-all" strategy, namely

$$\{\text{Final Result}\} = \arg(\max(o_j)) \ (j = 1, 2, \ldots, N_o) \ . \tag{3.15}$$

Then, the index number of the maximally activated output neuron generally indicates the final result.

3.2.4 Representing the Final Network Outputs by Kernel Memory

In both the cases of PNNs and GRNNs, unlike the general kernel memory concept, the activation of each Gaussian kernel in the hidden layer is directly transferred to the output neurons. However, in the kernel memory, this notion can also be altered, where appropriate, by modifying the manner of generating the activation values from the output neurons. (In the case of PNNs, from the structural point of view, this is already implied in terms of the topological equivalence property, as shown in the right part of Fig. 2.2.) This modification is possible, since, within the kernel memory context, the manner in calculating the network outputs is detached from the weight parameter tuning, unlike the conventional neural network principles. Thus, essentially, any function can be used to describe the network outputs, virtually with no numerical effect upon the memory storage. Moreover, such network outputs can even be forcibly represented by kernel units within the kernel memory concept. For instance, the output neurons within a PNN/GRNN o_j $(j = 1, 2, \ldots, N_o)$ in Fig. 3.3 can be represented in terms of the kernel units with a linear operation:

---

[5]In PNNs/GRNNs, the linear sum operators as defined in (3.14) may also be regarded as special forms of the kernel functions, where the inputs are the weighted version of $K_i(\mathbf{x})$.

$$o_j = K_j^o(\mathbf{y}) = \frac{1}{\xi}\mathbf{w}_j^T\mathbf{y} \qquad (3.16)$$

where $\mathbf{w}_j = [w_{1j}, w_{2j}, \ldots, w_{N_h,j}]^T$, ξ is a normalisation constant (given as that in (2.2) and (2.3)), and the vector comprising of the activations of the kernel units in the hidden layer $\mathbf{y} = [K_1^h(\mathbf{x}), K_2^h(\mathbf{x}), \ldots, K_{N_h}^h(\mathbf{x})]^T$ is now regarded as the input to the kernel unit K_j^o. As in the above, this principle can be applied to any modification of network outputs given hereafter.

Then, for the PNNs, the following simple modification to (3.14) can be alternatively made within the context of the topological equivalence property (see in Sect. 2.3):

$$o_j = f_2(\mathbf{x}) = \max(K_i^h(\mathbf{x})) \qquad (3.17)$$

where the output (kernel) o_j is regarded as the j-th sub-network output and the index i ($i = 1, 2, \ldots, N_j$, N_j: num. of kernels in Sub-network j) denotes the Gaussian kernel within the j-th sub-network.

However, unlike the case (3.14), since the above modification (3.17) is based upon only the *local* representation of the pattern space, it can be more effective to exploit both the *global* (i.e. (3.14)) and *local* (i.e. (3.17)) activations:

$$o_j = f_3(\mathbf{x}) = g(f_1(\mathbf{x}), f_2(\mathbf{x})) \qquad (3.18)$$

where $g(\cdot)$ is a certain function to yield a combination of the two factors, e.g. the convex mixture

$$g(x, y) = (1 - \lambda)x + \lambda y \qquad (3.19)$$

with $0 \leq \lambda \leq 1$. The factor λ may be determined *a priori* depending upon the application.

Similarly, we can also exploit the same strategy as in the k-nearest neighbours approach (see e.g. Duda et al., 2001), which may lead to a more consistent/robust result compared to the one given by (3.17); suppose that we have collected a total of K kernel units with maximal activations, the final result in (3.15) can be modified by taking a voting scheme amongst the first K kernel units:

1) Given the input vector $\mathbf{x}$, find the K kernel units with maximal activations $\check{K}_i$ ($i = 1, 2, \ldots, K$), amongst all the kernel units within the kernel memory. Initialise the variables $\rho_j = 0$ ($j = 1, 2, \ldots, N_o$).
2) Then, for $i = 1$ to K do:
3) If $\check{\eta}_i = j$ (i.e. the pattern data stored within the kernel unit $\check{K}_i$ falls in to Class j), update

$$\rho_j = \rho_j + 1 . \qquad (3.20)$$

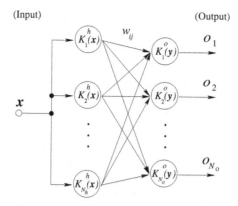

Fig. 3.4. A PNN/GRNN and its generalisation represented in terms of the kernel memory concept by using only kernel units; in the figure, the kernel functions $K_i^h(\mathbf{x})$ ($i = 1, 2, \ldots, N_h$, $\mathbf{x}$: the input vector) in the first (or hidden) layer are e.g. all Gaussian given by (3.13), whereas in the second (output) layer, the functions $K_j^o(\mathbf{y})$ ($j = 1, 2, \ldots, N_o$, $\mathbf{y} = [K_1^h(\mathbf{x}), K_2^h(\mathbf{x}), \ldots, K_{N_h}^h(\mathbf{x})]^T$) can be alternatively given by exploiting the representation such as (3.16) (i.e. for an ordinary PNN/GRNN), (3.17), (3.18), or (3.21)

4) Finally, the result is obtained by simply taking a maximum and is used as the output o_j:

$$o_j = \{\text{Final Result}\} = \max(\rho_j) \ (j = 1, 2, \ldots, N_o) \ . \qquad (3.21)$$

Note that all the modifications, i.e. (3.17), (3.18), and (3.21) given in the above can also be uniformly represented by kernel units as in (3.16) and can be eventually reduce to a simple kernel memory representation as shown in Fig. 3.4.

3.3 Topological Variations in Terms of Kernel Memory

In the previous section, it was described that both the neural network GRNN and PNN can be subsumed into the kernel memory concept, where only a layer of Gaussian kernels and a set of the kernels, each with a linear operator, are used, as shown in Fig. 3.4. However, within the kernel context, there essentially exist no such structural restrictions, and any topological form of the kernel memory representation is possible.

Here, we consider some topological variations in terms of the kernel memory.

3.3.1 Kernel Memory Representations
for Multi-Domain Data Processing

The kernel memory in Fig. 3.3 or 3.4 can be regarded as a single-input multi-output (SIMO) (more appropriately, a single-*domain*-input multi-output

(SDIMO) system) in that only a single (domain) input vector ($\mathbf{x}$) and multiple outputs (i.e. N_o outputs) are used.

In contrast, the kernel memory shown in Fig. 3.5[6] can be viewed as a multi-input multi-output (MIMO)[7] (i.e. a three-input three-output) system, since, in this example, three different domain input vectors $\mathbf{x}^m = [x^m(1), x^m(2), \ldots, x^m(N_m)]^T$ ($m = 1, 2, 3$, and the length N_m of the input vector $\mathbf{x}^m$ can be varied) and the three output kernel units are used.

In the figure, $K_i^m(\mathbf{x}^m)$ denotes the i-th kernel which is responsible for the m-th domain input vector $\mathbf{x}^m$ and the mono-directional connections between the kernel units and output kernels (or, unlike the original PNN/GRNN, the bi-directional connections between the kernels) represent the link weights w_{ij}. Note that, as well as for clarity (see the footnote[6]), the three output kernel units, $K_1^o(\mathbf{y})$, $K_2^o(\mathbf{y})$, and $K_3^o(\mathbf{y})$, the respective kernel functions of which are defined as the final network outputs, are considered. As the output kernel units for the PNN/GRNN in Fig. 3.4, the input vector $\mathbf{y}$ of the output kernel K_j^o ($j = 1, 2, 3$) in Fig. 3.5 is given by a certain function which takes into account e.g. the transfers of the activation from the respective kernel units so connected in the previous layer, i.e. $K_i^m(\mathbf{x}^m)$, via the link weights w_{ij}.

Note also that, hereafter, in order to distinguish the two types of connections (or the links) between the nodes within the network structure in terms of the kernel memory concept, two different colours will be used as in Fig. 3.5; the connection in *grey* line denotes the ordinary "link weight" (i.e. the link with a weighting factor), whereas that in *grey* indicates either the *input* to or activation from the kernel unit (i.e. *output*, due to the kernel function) which is normally represented without such weighting factor.

Two Ways of Forwarding Data to a Kernel Unit

Then, as in the structures in Figs. 3.4 and 3.5, it is considered that there are two manners of forwarding the data to a single kernel in terms of the kernel unit representation shown in Fig. 3.1/3.2:

---

[6]The kernel memory structure depicted in Fig.3.5 exploits the modified kernel unit representation shown in Fig. 3.2, instead of the original as in Fig. 3.1. In Fig. 3.5, the three output kernels, K_1^o, K_2^o, and K_3^o, thus represent the nodes indicating class labels. As discussed in Sect. 3.2.2, this representation can be more convenient to depict the topological structure.

[7]Alternatively, this can be called as a multi-*domain*-input multi-output (MDIMO) system.

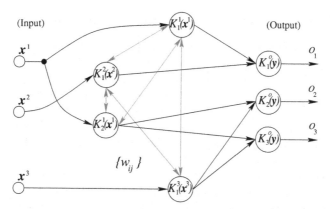

Fig. 3.5. Example 1 – a multi-input multi-output (MIMO) (or, a three-input three-output) system in terms of kernel memory; in the figure, it is considered that there are three modality-dependent inputs $\mathbf{x}^m = [x^m(1), x^m(2), \ldots, x^m(N_m)]$, ($m = 1, 2, 3$) to the MIMO system and that four kernel units K_i^m ($i = 1, 2$) to process the modality-dependent inputs and three output kernels K_1^o, K_2^o, and K_3^o. Note that, as in this example, it is possible that the network structure is not necessarily fully-connected, whilst allowing the lateral connections between the kernel units, within the kernel memory principle

The input data giving as

1. The data input to the kernel itself;
2. The transfer of the activation from other connected kernel(s) via the link weight(s) w_{ij}, by exploiting the pointers to other kernel units so attached p_j ($j = 1, 2, \ldots, N_p$)

For example, the kernel units such as $K_1^1(\mathbf{x}^1)$ and $K_1^2(\mathbf{x}^2)$ in Fig. 3.5 have both the two types of the input data, whilst the kernel units $K_j^o(\mathbf{y})$ representing the respective final network outputs can be only activated by transfer of the activation from the other non-output kernels. (Note that the former case always yields mono-directional connections.) In the early part of the next chapter, we will consider how actually the two ways of activation from a single kernel unit as in the above can be modelled.

Now, to see how the MIMO system in Fig. 3.5 works, consider the situation where the input vector $\mathbf{x}^1$ is given as the feature vector obtained from the voice sound uttered by a specific person which activates the kernel K_1^1. Then, since the kernel K_1^1 is connected to K_1^3 via the bi-directional link weight in between, it is possible to design the system such that, without the direct excitation by the feature vector $\mathbf{x}^3$ obtained from, say, the facial image of the corresponding person, the kernel K_1^3 can be subsequently activated, due to the transfer of the activation from the kernel unit K_1^1. However, these subsequent activations can occur, only when K_1^1 is excited, not by the activation of the kernel in the same domain K_2^1, since K_1^3 is not linked to K_2^1, as in Fig. 3.5.

Comparison with Conventional Modular Approaches

In general, it is said that the network structure in Fig. 3.5 acts as an integrated pattern classifier and can process the input patterns in different domains simultaneously/ in parallel; *even without having the input in a particular domain, the kernel can be excited by transfer of the activations from other kernel units*. Moreover, within the kernel memory concept, it is possible for the structure of the kernel memory not always to be fully-connected. These features have not generally been considered within the traditional neural network context.

In addition, such features cannot be easily realised by simply considering a mixture of the pattern classifiers (or agents), each of which is responsible for the classification task in a particular domain, as in typical modular approaches (see e.g. Haykin, 1994). This is since they normally exploit conventional neural network architectures (for the applications to sensor fusion, see e.g. Wolff et al., 1993; Colla et al., 1998), in which all the nodes are usually fully-connected (without allowing the lateral (but not necessarily for all the nodes) connections between different domains), and function (only) when *all* the input data are presented (at a time) to the hybrid system. Moreover, in respect to conventional neural architectures, the kernel memory is more advantageous in that

1) The structure of each network in the modular approach is usually more complex than a single kernel unit;
2) Long iterative training is generally needed to make each agent work properly (and therefore time consuming);
3) The question as to how to control such agents in a uniform and/or efficient manner remains (typically, another network must be trained, which is often called a "gating" network (see e.g. Haykin, 1994)).

Representing Directional Links

In the previous topological representation as shown in Fig. 3.5, it has been described that the way of transferring data amongst kernel units can be classified into two types. In this subsection, before moving on to other topologies, we consider a little further the directed links, i.e. bi-/mono-directional data transmissions.

As in Fig. 3.5, the connection such as that between $K_1^1(\mathbf{x}^1)$ and $K_1^2(\mathbf{x}^2)$ is established via a bi-directional link, whilst that between $K_1^1(\mathbf{x}^1)$ and the kernel K_1^o is via a mono-directional link. Then, it is considered that a mono-directional link can be the representation of an excitatory/inhibitory synapse[8], in the neurophysiological context (see e.g. Koch, 1999), and implemented by means of electronic devices such as diodes or transistors.

---

[8]In this book, unlike ordinary neural network schemes, both the excitatory and inhibitory synapses are considered to be represented in terms of directed graphs. However, it is straightforward to return from such directed graphs to ordinary schemes (i.e. the excitatory synapses are represented by positive weighted values, whilst the inhibitory are by the negative values).

Thus, each bi-directional link in Fig. 3.5 may be composed of a pair of mono-directional links in which the directions are opposite to each other, with a different weighting setting in between. However, for convenience, we hereafter regard the notation of the link weight between a pair of kernel units K_A and K_B w_{AB} as simply the unique weight value in between, i.e.

$$w_{AB} = w_{BA} \qquad (3.22)$$

unless denoted otherwise; only the arrow(s) represents the directional flow(s)[9].

A Bi-directional Representation

Figure 3.6 illustrates another example, where there are only three kernels[10] but their roles are all different; kernel K_1 is responsible for sound input, whereas K_2 is for image input, as in the previous example, and both K_1 and K_2 are connected to K_3, the kernel of which integrates i.e. the transfer of the activation from both K_1 and K_2.

In this example, it is considered that either i) the input vector $\mathbf{x}$ of the kernel unit K_3 is given as $\mathbf{x} = [w_{13} \ w_{23}]^T$, instead of the feature vector obtained from the ordinary input, or ii) there is no input vector directly given to K_3 but, rather, the kernel K_3 can be activated by transfer via the link weights i.e. w_{13} and/or w_{23} (then, apparently, the representation in Fig. 3.6 implies the latter case).

Note that in this example there are no explicit output kernels given as in Fig. 3.5, since the functionality of this kernel memory is different: consider the case where a particular feature given by $\mathbf{x}^1$ activates the kernel K_1 and where K_2 is simultaneously/in parallel activated by $\mathbf{x}^2$. This is similar to the situation where both auditory and visual information are simultaneously given to the memory. Then, provided that we choose a Gaussian kernel function for all the three kernel units and that the input vector $\mathbf{x}$ is sufficiently close to the centroid vector to excite the kernel K_i ($i = 1, 2, 3$), we can make this kernel memory network also eventually output the centroid vector $\mathbf{c}_i$, apart from the ordinary output values obtained as the activation of the respective kernel functions, and, eventually, the activation of the kernel K_3 is furtherly transferred to other kernel(s) via the link weight w_{3k} ($k \neq 1, 2, 3$). In such a situation, it is considered that the kernel K_3 integrates the information transferred from both K_1 and K_2 and hence imitates the concept or "Gestalt"

---

[9]As the directed graphs in general graph theory (see e.g. Christofides, 1975), where appropriate, we may alternatively consider both the link weights w_{AB} and w_{BA}, in order to differentiate the weight value with respect to the direction. Moreover, hereafter, the link connections without arrows represent bi-directional flows, which satisfy the relation in (3.22), unless denoted otherwise.

[10]In the figure, both the superscripts for the input vectors $\mathbf{x}$ indicating the domain numbers and the input argument of the kernel units are omitted for clarity.

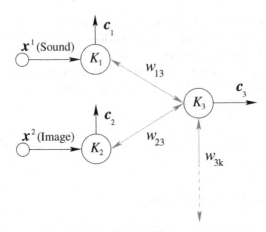

Fig. 3.6. Example 2 – a bi-directional MIMO system represented by kernel memory; in the figure, each of the three kernel units receives and yields the outputs, representing the bi-directional flows. For instance, when both the two modality-dependent inputs $\mathbf{x}^1$ and $\mathbf{x}^2$ are simultaneously presented to the kernel units K_1 and K_2, respectively, K_3 may be subsequently activated via the transfer of the activations from K_1 and K_2, due to the link weight connections in between (thus, *feedforward*). In reverse, the excitation of the kernel unit K_3 can cause the subsequent activations from K_1 and K_2 via the link weights w_{12} and w_{13} (i.e. *feedback*). Note that, instead of ordinary outputs, each kernel is considered to output its template (centroid) vector in the figure

formation (i.e. related to the *concept formation*; to be described in Chap. 9). Thus, the information flow in this case is *feedforward*:

$$\mathbf{x}^1, \mathbf{x}^2 \to K_1, K_2 \to K_3 .$$

In contrast, if such a "Gestalt" kernel K_3 is (somehow) activated by the other kernel(s) via w_{3k} and the activation is transferred back to both kernels K_1 and K_2 via the respective links w_{13} and w_{23}, the information flow is, in turn, *feedback*, since

$$w_{3k} \to K_3 \to K_1, K_2 .$$

Therefore, the kernel memory as in Fig. 3.6 represents a *bi-directional* MIMO system.

As a result, it is also possible to design the kernel memory in such a way that the kernels K_1 and K_2 eventually output the centroid vector $\mathbf{c}_1$ and $\mathbf{c}_2$, respectively, and if the appropriate decoding mechanisms for $\mathbf{c}_1$ and $\mathbf{c}_2$ are given (as external devices), we could even restore the complete information (i.e. in this example, this imitates the mental process to remember both the sound and facial image of a specific person at once).

Note that both the MIMO systems in Figs. 3.5 and 3.6 can in principle be viewed as graph theoretic networks (see e.g. Christofides, 1975) and the

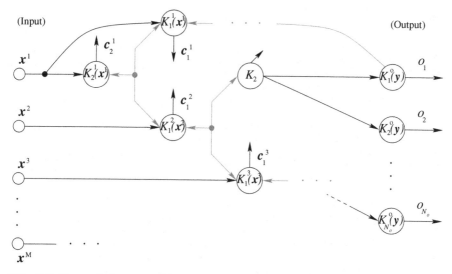

Fig. 3.7. Example 3 – a tree-like representation in terms of a MIMO kernel memory system; in the figure, it can be considered that the kernel unit K_2 plays a role for the concept formation, since the kernel does not have any modality-dependent inputs

detailed discussion of how such directional flows can be realised in terms of kernel memory is left to the later subsection "3) Variation in Generating Outputs from Kernel Memory: Regulating the Duration of Kernel Activations" (in Sect. 3.3.3).

Other Representations

The bi-directional representation as in Fig. 3.6 can be regarded as a simple model of concept formation (to be described in Chap. 9), since it can be seen that the kernel network is an integrated incoming data processor as well as a composite (or associative) memory. Thus, by exploiting this scheme, more sophisticated structures such as the tree-like representation in Fig. 3.7, which could be used to construct the systems in place of the conventional symbol-based database, or lattice-like representation in Fig. 3.8, which could model the functionality of the retina, are possible. (Note that, the kernel K_2 illustrated around in the centre of Fig. 3.7, does not have the ordinary modality-dependent inputs, i.e. $\mathbf{x}^i$ $(i = 1, 2, \ldots, M)$, as this kernel plays a role for the concept formation (in Chap. 9), similar to the kernel K_3 in Fig. 3.6.)

3.3.2 Kernel Memory Representations for Temporal Data Processing

In the previous subsection a variant of network representations in terms of kernel memory has been given. However, this has not taken into account the

(Input)   (Output)

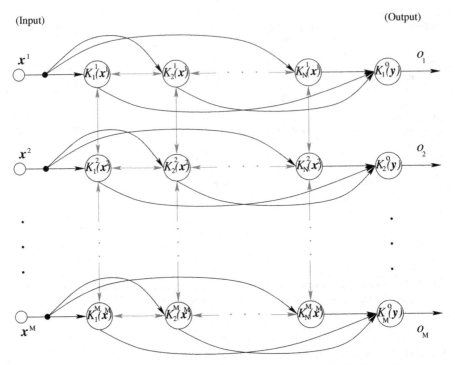

Fig. 3.8. Example 4 – a lattice-like representation in terms of MIMO kernel memory system

functionality of temporal data processing. Here, we consider another variation of the kernel memory model within the context of temporal data processing.

In general, the connectionist architectures as used in pattern classification tasks take only *static* data into consideration, whereas the time delay neural network (TDNN) (Lang and Hinton, 1988; Waibel, 1989) or, in a wider sense of connectionist models, the adaptive filters (ADFs) (see e.g. Haykin, 1996) concern the situations where both the input pattern and corresponding output are varying in time. However, since they still resort to a gradient-descent type algorithm such as least mean square (LMS) or BP for parameter estimation, a flexible reconfiguration of the network structure is normally very hard, unlike the kernel memory approach.

Now, let us turn back to temporal data processing in terms of kernel memory: suppose that we have collected a set of single domain inputs[11] obtained during the period of (discrete) time P (written in a matrix form):

$$\mathbf{X}(n) = [\mathbf{x}(n), \mathbf{x}(n-1), \ldots, \mathbf{x}(n-P+1)]^T \qquad (3.23)$$

where $\mathbf{x}(n) = [x_1(n), x_2(n), \ldots, x_N(n)]^T$. Then, considering the temporal variations, we may use a matrix form, instead of vector, within the template data

---

[11]The extension to multi-domain inputs is straightforward.

stored in each kernel, and, if we choose a Gaussian kernel , it is normally convenient to regard the template data in the form of a *template matrix* (or *centroid matrix* in the case of a Gaussian response function) $\mathbf{T} \in \Re^{N \times P}$, which covers the period of time P:

$$
\mathbf{T} = \begin{bmatrix} \mathbf{t}_1 \\ \mathbf{t}_2 \\ \vdots \\ \mathbf{t}_N \end{bmatrix} = \begin{bmatrix} t_1(1) & t_1(2) & \dots & t_1(P) \\ t_2(1) & t_2(2) & \dots & t_2(P) \\ \vdots & \vdots & \ddots & \vdots \\ t_N(1) & t_N(2) & \dots & t_N(P) \end{bmatrix} \tag{3.24}
$$

where the column vectors contain the temporal data at the respective time instances up to the period P.

Then, it is straightforward to generalise the kernel memory that employs both the properties of temporal and multi-domain data processing.

3.3.3 Further Modification of the Final Kernel Memory Network Outputs

With the modifications of the temporal data processing as described in Sect. 3.3.2, we may accordingly redefine the final outputs from kernel memory. Although many such variations can be devised, we consider three final output representations which are considered to be helpful in practice and can be exploited e.g. for describing the notions related to mind in later chapters.

1) Variation in Generating Outputs from Kernel Memory: Temporal Vector Representation

One of the final output representations can be given as a *time sequence* of the outputs:

$$
\mathbf{o}_j(n) = [o_j(n), o_j(n-1), \dots, o_j(n - \check{P} + 1)]^T \tag{3.25}
$$

where each output is now given in a vector form as $\mathbf{o}_j(n)$ $(j = 1, 2, \dots, N_o)$ (instead of the scalar output as in Sect. 3.2.4) and $\check{P} \leq P$. This representation implies that not all the output values obtained during the period P are necessarily used, but partially, and that the output generation(s) can be *asynchronous* (in time) to the presentation of the inputs to the kernel memory. In other words, unlike conventional neural network architectures, the *timing* of the final output generation from kernel memory may differ from that of the input presentation, within the kernel memory context.

Then, each element in the output vector $\mathbf{o}_j(n)$ can be given, e.g.

$$
o_j(n) = \text{sort}(\max(\theta_{ij}(n))) \tag{3.26}
$$

where the function $\text{sort}(\cdot)$ returns the multiple values given to the function sorted in a descending order, i denotes the indices of all the kernels within a specific region(s)/the entire kernel memory, and

$$\theta_{ij}(n) = w_{ij}K_i(\mathbf{x}(n)) \ . \tag{3.27}$$

The above variation in (3.26) does not follow the ordinary "winner-takes-all" strategy but rather yields multiple output candidates which could, for example, be exploited for some more sophisticated decision-making processing (i.e. this is also related to the topic of *thinking*; to be described later in Chaps. 7 and 9).

2) Variation in Generating Outputs from Kernel Memory: Sigmoidal Representation

In contrast to the vector form in (3.25), the following scalar output o_j can also be alternatively used within the kernel memory context:

$$o_j(n) = f(\boldsymbol{\theta}_{ij}(n)) \tag{3.28}$$

where the activations of the kernels within a certain region(s)/the entire memory $\boldsymbol{\theta}_{ij}(n) = [\theta_{ij}(n), \theta_{ij}(n-1), \dots, \theta_{ij}(n-P+1)]^T$ and the cumulative function $f(\cdot)$ is given in a sigmoidal (or "squash") form, i.e.

$$f(\boldsymbol{\theta}_{ij}(n)) = \frac{1}{1 + \exp(-b\sum_{k=0}^{P-1}\theta_{ij}(n-k))} \tag{3.29}$$

where the coefficient b determines the steepness of the sigmoidal slope.

An Illustrative Example of Temporal Processing – Representation of Spike Trains in Terms of Kernel Memory

Note that, by exploiting the output variations given in (3.25) or (3.29), it is possible to realise the kernel memory which can be alternative to the TDNN (Lang and Hinton, 1988; Waibel, 1989) or the pulsed neural network (Dayhoff and Gerstein, 1983) models, with much more straightforward and flexible reconfiguration property of the memory/network structures.

   As an illustrative example, consider the case where a sparse template matrix $\mathbf{T}$ of the form (3.24) is used with the size of (13×2), where the two column vectors $\mathbf{t}_1$ and $\mathbf{t}_2$ are given as

$$\mathbf{t}_1 = [2\ 0\ 0\ 0\ 0.5\ 0\ 0\ 0\ 1\ 0\ 0\ 0\ 1]$$
$$\mathbf{t}_2 = [2\ 1\ 2\ 0\ 0\ 0\ 0\ 0\ 1\ 0.5\ 1\ 0\ 0]\ ,$$

i.e. the sequential values in the two vectors depicted in Fig. 3.9 can be used to represent the situation where a cellular structure gathers for the period of time $P(= 13)$ and then stores the patterns of spike trains coming from other neurons (or cells) with different *firing rates* (see e.g. Koch, 1999).

   Then, for instance, if we choose a Gaussian kernel and the overall synaptic inputs arriving at the kernel memory *match* the stored spike pattern to

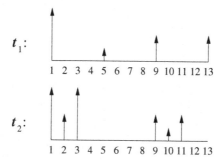

Fig. 3.9. An illustrative example: representing the spike trains in terms of the sparse template matrix of a kernel unit for temporal data processing (where each of the two vectors in the template matrix contains a total of 13 spikes)

a certain extent (i.e. determined by both the threshold θ_K and radius σ, as described earlier), the overall excitation of the cellular structure (in terms of the activation from a kernel unit) can occur due to the stimulus and subsequently emit a spike (or train) from itself.

Thus, the pattern matching process of the spike trains can be modelled using a sliding window approach as in Fig. 3.10; the spike trains stored within a kernel unit in terms of a sparse template (centroid) matrix are compared with the input patterns $\mathbf{X}(n) = [\mathbf{x}_1(n)\ \mathbf{x}_2(n)]$ at each time instance n.

3) Variation in Generating Outputs from Kernel Memory: Regulating the Duration of Kernel Activations

The third variation in generating the outputs from kernel memory is due to the introduction of the decaying factor for the duration of kernel excitations. For the output generation of the i-th kernel, the following modification can be considered:

$$K_i(\mathbf{x}, n_i) = \exp(-\kappa_i n_i)K_i(\mathbf{x}) \tag{3.30}$$

where n_i[12] denotes the time index for describing the decaying activation of K_i and the duration of the i-th kernel output is regulated by the newly introduced factor κ_i, which is hereafter called *activation regularisation factor*. (Note that the time index n_i is used independent of the kernels, instead of the unique index n, for clarity.) Then, the variation in (3.30) indicates that the activation of the kernel output can be decayed in time.

In (3.30), the time index n_i is reset to zero, when the kernel K_i is activated after a certain interval from the last series of activations, i.e. the period of time when the following relation is satisfied (i.e. the counter relation in (3.12)):

$$K_i(\mathbf{x}_i, n_i) < \theta_K \tag{3.31}$$

---

[12]Without loss of generality, here the time index n_i is again assumed to be discrete; the extension to continuous time representation is straightforward.

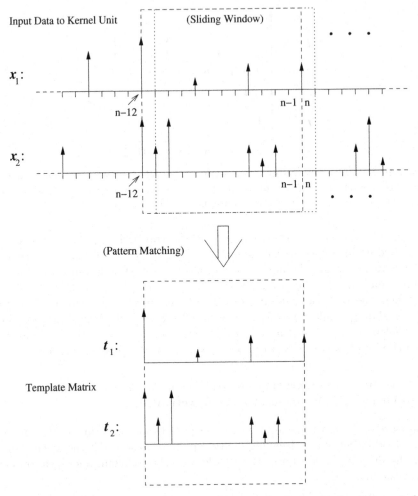

Fig. 3.10. Illustration of the pattern matching process in terms of a sliding window approach. The spike trains stored within a kernel unit in terms of a sparse template matrix are compared with the current input patterns $\mathbf{X}(n) = [\mathbf{x}_1(n)\ \mathbf{x}_2(n)]$ at each time instance n

3.3.4 Representation of the Kernel Unit Activated by a Specific Directional Flow

In the previous examples of the MIMO systems as shown in Figs. 3.5–3.8, some of the kernel units have (mono-/bi-)directional connections in between. Here, we consider the kernel unit that can be activated when a specific directional flow occurs between a pair of kernel units, by exploiting both the notation of the template matrix as given in (3.24) and modified output in (3.30) (the fundamental principle of which is motivated by the idea in Kinoshita (1996)).

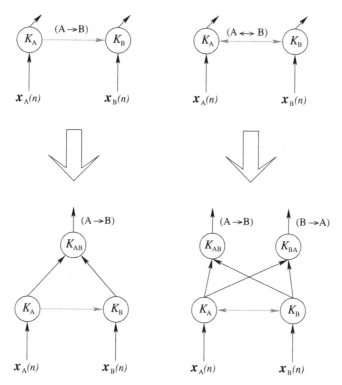

Fig. 3.11. Illustration of both the mono- (on the left hand side) and bi-directional connections (on the right hand side) between a pair of kernel units K_A and K_B (cf. the representation in Kinoshita (1996) on page 97); in the lower part of the figure, two additional kernel units K_{AB} and K_{BA} are introduced to represent the respective directional flows (i.e. the kernel units that *detect* the transfer of the activation from one kernel unit to the other): $K_A \rightarrow K_B$ and $K_B \rightarrow K_A$

Fig. 3.11 depicts both the mono- (on the left hand side) and bi-directional connections (on the right hand side) between a pair of kernel units K_A and K_B (cf. the representation in Kinoshita (1996) on page 97).

In the lower part of the figure, two additional kernel units K_{AB} and K_{BA} are introduced to represent the respective directional flows (i.e. the kernel units that *detect* the transfer of the activation from one kernel unit to the other): $K_A \rightarrow K_B$ and $K_B \rightarrow K_A$.

Now, let us firstly consider the case where the template matrix of both the kernel units K_{AB} and K_{BA} is composed by the series of activations from the two kernel units K_A and K_B, i.e.:

$$\mathbf{T}_{AB/BA} = \begin{bmatrix} t_A(1) \ t_A(2) \ldots t_A(p) \\ t_B(1) \ t_B(2) \ldots t_B(p) \end{bmatrix} \qquad (3.32)$$

where p represents the number of the activation status from time n to $n-p+1$ to be stored in the template matrix and the element $t_i(j)$ (i: A or B, $j = 1, 2, \ldots, p$) can be represented using the modified output given in (3.30) as[13]

$$t_i(j) = K_i(\mathbf{x}_i, n - j + 1) , \qquad (3.33)$$

or, alternatively, the indicator function

$$t_i(j) = \begin{cases} 1 \; ; \; \text{if } K_i(\mathbf{x}_i, n - j + 1) \geq \theta_K \\ 0 \; ; \; \text{otherwise} \end{cases} \qquad (3.34)$$

(which can also represent a collection of the spike trains from two neurons.)

Second, let us consider the situation where the activation regularisation factor of one kernel unit K_A, say, κ_A satisfies the relation:

$$\kappa_A < \kappa_B \qquad (3.35)$$

so that, at time n, the kernel K_B is not activated, whereas the activation of K_A is still maintained. Namely, the following relations can be drawn in such a situation:

$$K_A(\mathbf{x}_A(n - p_d + 1)) \; , \; K_B(\mathbf{x}_B(n - p_d + 1)) \geq \theta_K$$

$$K_A(\mathbf{x}_A(n)) \geq \theta_K$$
$$K_B(\mathbf{x}_B(n)) < \theta_K \qquad (3.36)$$

where p_d is a positive value. (Nevertheless, due to the relation (3.35) above, it is considered that the decay in the activation of both the kernel units K_A and K_B starts to occur at time n, given the input data.) Figure 3.12 illustrates an example of the regularisation factor setting of the two kernel units K_A and K_B as in the above and the time-wise decaying curves. (In the figure, it is assumed that $p_d = 4$ and $\theta_K = 0.7$.)

Then, if $p_d < p$, and, using the representation of the indicator function given by (3.34), for instance, the matrix

$$\mathbf{T}_{AB} = \begin{bmatrix} 0 \; 1 \; 1 \; 1 \; 1 \; 0 \\ 0 \; 0 \; 1 \; 1 \; 1 \; 1 \end{bmatrix} \qquad (3.37)$$

can represent the template matrix for the kernel unit K_{AB} (i.e. in this case, $p = 6$ and $p_d = 4$) and hence the directional flow of $K_A \to K_B$, since the matrix representation describes the following *asynchronous* activation pattern between K_A and K_B:

1) At time $n - 5$, neither K_A nor K_B is activated;
2) At time $n - 4$, the kernel unit K_A is activated (but not K_B);

---

[13]Here, for convenience, a unique time index n is considered for all the kernels in Fig. 3.11, without loss of generality.

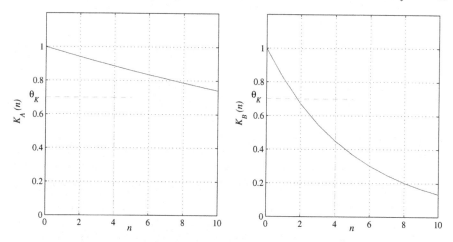

Fig. 3.12. Illustration of the decaying curves $\exp(-\kappa_i \times n)$ (i: A or B) for modelling the time-wise decaying activation of the kernel units K_A and K_B; $\kappa_A = 0.03$, $\kappa_B = 0.2$, $p_d = 4$, and $\theta_K = 0.7$

3) At time $n - 3$, the kernel unit K_B is then activated;
4) The activation of both the kernel units K_A and K_B lasts till the time $n - 1$;
5) Eventually, due to the presence of the decaying factor κ_B, the kernel unit K_B is not activated at time n.

In contrast to (3.37), the matrix (with inverting the two row vectors in (3.37))

$$\mathbf{T}_{BA} = \begin{bmatrix} 0\ 0\ 1\ 1\ 1\ 1 \\ 0\ 1\ 1\ 1\ 1\ 0 \end{bmatrix} \tag{3.38}$$

represents the directional flow of $K_B \to K_A$ and thus the template matrix of K_{BA}.

Therefore, provided a Gaussian response function (with appropriately given the radius, as defined in (3.8)) is selected for either the kernel unit K_{AB} or K_{BA}, if the kernel unit receives a series of the lasting activations from K_A and K_B as the inputs (i.e. represented in spiky trains), and the activation patterns are close to those stored as in (3.37) or (3.38), the kernel units can represent the respective directional flows.

A Learning Strategy to Obtain the Template Matrix for Temporal Representation

When the asynchronous activation between K_A and K_B occurs and provided that $p = 3$ (i.e. for the kernel unit K_{AB}/K_{BA}), one of the following patterns can be obtained using the indicator function representation of the spike trains by (3.34):

$$K_A(\mathbf{x}_A(n)): \cdots 0\boxed{1\ 0\ 0}0\ 0\ 0 \cdots$$
$$K_B(\mathbf{x}_B(n)): \cdots 0\boxed{0\ 0\ 0}0\ 1\ 0 \cdots$$

In the above, it is not sufficient to represent the asynchronous activation pattern by K_{AB} (or K_{BA}).

It is then considered that there are two alternative ways to adjust the template matrix for the kernel unit K_{AB} (or K_{BA}) that can represent the asynchronous activation pattern between the kernel units K_A and K_B:

1. Adjust the size of the template matrix $\mathbf{T}_{AB}$ (i.e. varying the factor p; in this case, assuming that $\kappa_i = \kappa_{init}$ $(\forall i))$;
2. Update the activation regularisation factors for both the kernel units K_A and K_B

For the former, if we increase the number of columns of the template matrix p, until the activation from K_A and K_B appears in both the rows (i.e. $p = 5$):

$$K_A(\mathbf{x}_A(n)): \cdots 0\boxed{1\ 0\ 0\ 0\ 0}0 \cdots$$
$$K_B(\mathbf{x}_B(n)): \cdots 0\boxed{0\ 0\ 0\ 0\ 1}0 \cdots$$

the asynchronous activation pattern can be represented by the template matrix, i.e.

$$\mathbf{T}_{AB} = \begin{bmatrix} 1\ 0\ 0\ 0\ 0 \\ 0\ 0\ 0\ 0\ 1 \end{bmatrix} \tag{3.39}$$

An Alternative Learning Scheme – Updating the Activation Regularisation Factors

Alternatively, the asynchronous activation pattern between K_A and K_B can be represented by updating the activation regularisation factors for both the kernel unit K_A and K_B, without varying p: provided that the regularisation factor for all the kernel units are initially set as $\kappa_i = \kappa_{init}$ (where κ_{init} is a certain positive constant), we update the activation regularisation factors for both the kernel unit K_A and K_B, i.e. κ_A and κ_B. Then, we may resort to the following updating rule:

[Updating Rule for the Activation Regularisation Factor κ_i]

1) Initially, set $\kappa_i = \kappa_{init}$ $(\forall i)$.

2)

- For a certain period of time, if the kernel unit K_i has activated repetitively, update its regularisation factor κ_i as:

$$\kappa_i = \begin{cases} \kappa_i - \delta\kappa_1 & \text{; if } \kappa_i > \kappa_{min} \\ \kappa_{min} & \text{; otherwise} \end{cases} \qquad (3.40)$$

- In contrast, for a certain period of time, if there is no activation from K_i, increase the value of κ_i:

$$\kappa_i = \kappa_i + \delta\kappa_2 \qquad (3.41)$$

---

in the above where $\kappa_{min}(\geq 0)$ is the minimum value for the regularisation factor, and $\delta\kappa_1$ and $\delta\kappa_2$ are its decremental and incremental adjustment factor, respectively.

For instance, if the duration of activation from K_A becomes longer and, accordingly, if we successfully obtain the following pattern using the scheme similar to the above

$$K_A(\mathbf{x}_A(n)): \cdots 0\ 1\ 1\ \boxed{1\ 0\ 0}\ 0 \cdots$$
$$K_B(\mathbf{x}_B(n)): \cdots 0\ 0\ 0\ \boxed{0\ 0\ 1}\ 0 \cdots$$

the template matrix (i.e. $p = 3$)

$$\mathbf{T}_{AB} = \begin{bmatrix} 1 & 0 & 0 \\ 0 & 0 & 1 \end{bmatrix} \qquad (3.42)$$

can represent the asynchronous activation of $K_A \rightarrow K_B$.

The above scheme can be applied, under the assumption that the duration of activation K_A can be different from that of K_B by varying κ_A/κ_B.

Nevertheless, the directed conections also have to be established within the context of general learning (to be described in Chap. 7). In later chapters, it will then be discussed how the principle of the directed connections between the kernel units is exploited further and can significantly enhance the utility for modelling various notions related to artificial mind system, e.g. the thinking, language, and the semantic networks/lexicon module.

3.4 Chapter Summary

In this chapter, a novel kernel memory concept has been described, which can subsume conventional connectionist principles.

The fundamental principle of kernel memory concept is pretty simple; the kernel memory comprises of multiple kernel units and their link weights which only represent the strengths of the connections in between. Within the kernel memory principle, the following three types of kernel units have been considered:

1) A kernel unit which has both the input and template vector (i.e. the centroid vector in the case of a Gaussian kernel function) and generates the output, according to the similarity of the two vectors. (However, as described in the next chapter, it is also possible to consider the case where the activation can be due to the transfer of the activations from other kernel units connected via the link weights, as given by (4.3) or (4.4), to be described later).

2) A kernel unit functioning similar to the above, except that the input vector is merely composed of the activations from other kernel units (i.e. as the neurons in the conventional ANNs). (However, for this type, it still is possible that the input vector consists of both the activations from other kernels and the regular input data.)

3) A kernel unit which represents a symbolic node (as in the conventional connectionist model, or the one with a kernel function given by (3.11)). This sort of kernel unit is useful in practice, e.g. to investigate the intermediate / internal states of the kernel memory. In pattern recognition problems, for instance, these nodes are exploited to tell us the recognition results. This issue will be furtherly discussed within a more global context of *target responses* in Chap. 7 (Sect. 7.5).

Then, within the kernel memory concept, any rule can be developed to establish the link weight connections between a pair of kernel units, without directly affecting the contents of the memory.

In the next chapter, as a pragmatic example, the properties of the kernel memory concept are exploited to develop a self-organising network model, and we will see how such a kernel network behaves.

4

The Self-Organising Kernel Memory (SOKM)

4.1 Perspective

In the previous chapter, various topological representations in terms of the kernel memory concept have been discussed together with some illustrative examples. In this chapter, a novel unsupervised algorithm to train the link weights between the KFs is given by extending the original Hebb's neuropsychological concept, whereby the self-organising kernel memory (SOKM)[1] is proposed.

The link weight adjustment algorithm does not involve any gradient-descent type numerical approximation (or the so-called "delta rule") as in the conventional approaches, but simply varies the strength of the connections between KFs according to their activations. Thus, in terms of the SOKM, any topological representation of the data structure is possible, without suffering from any numerical instability problems. Moreover, the activation of a particular node (i.e. the KF) is conveyed to the other nodes (if any) via such connections. Then, this manner of data transfer represents more life-like/cybernetic memory. In the SOKM context, each kernel unit is thus regarded as a new memory element, which can at the same time exhibit the generalisation capability, instead of the ordinary node as used in conventional connectionist models.

---

[1]Here, unlike the ordinary self-organising maps (SOFMs) (Kohonen, 1997), the utility of the term "self-organising" also implies "construction" in the sense that the kernel memory is constructed from scratch (i.e. without any nodes; from a blank slate (Platt, 1991)). In the SOFMs, the utility is rather limited; all the nodes are already located in a fixed two-dimensional space and the clusters of nodes are formed in a self-organising manner within the fixed map, whilst both the size/shape of the entire network (i.e. the number of nodes) and the number/manner of connections are dynamically changed within the SOKM principle.

Tetsuya Hoya: *Artificial Mind System – Kernel Memory Approach*, Studies in Computational Intelligence (SCI) **1**, 59–80 (2005)
www.springerlink.com

4.2 The Link Weight Update Algorithm (Hoya, 2004a)

In Hebb (1949) (p.62), Hebb postulated, *"When an axon of cell A is near enough to excite a cell B and repeatedly or persistently takes part in firing it, some growth process or metabolic change takes place in one or both cells such that A's efficiency, as one of the cells firing B, is increased."*

In the SOKM, the "link weights" (or simply, "weights") between the kernels are defined in this neuropsychological context. Namely, the following conjecture can be firstly drawn:

Conjecture 1: When a pair of kernels K_i and K_j ($i \neq j$, $i, j \in \{$all indices of the kernels$\}$) in the SOKM are excited repeatedly, a new link weight w_{ij} between K_i and K_j is formed. Then, if this occurs intermittently, the value of the link weight w_{ij} is increased.

In the above, Hebb's original postulate for the adjacent locations of cell A and B is not considered; since, in actual hardware implementation of the proposed scheme (e.g. within the memory system of a robot), it may not always be crucial for such place adjustment of the kernels. Secondly, Hebb's postulate implies that the excitation of cell A may occur due to the *transfer* of activations from other cells via the synaptic connections. This can lead to the following conjecture:

Conjecture 2: When a kernel K_i is excited and one of the link weights is connected to the kernel K_j, the excitation of K_i is transferred to K_j via the link weight w_{ij}. However, the amount of excitation depends upon the (current) value of the link weight.

4.2.1 An Algorithm for Updating Link Weights Between the Kernels

Based upon **Conjectures 1 and 2** above, the following algorithm for updating the link weights between the kernels is given:

[The Link Weight Update Algorithm]

1) If the link weight w_{ij} is already established, decrease the value according to:

$$w_{ij} = w_{ij} \times \exp(-\xi_{ij}) \qquad (4.1)$$

---

2) If the simultaneous excitation of a pair of kernels K_i and K_j $(i \neq j)$ occurs (i.e. when the activation is above a given threshold as in (3.12); $K_i \geq \theta_K$) and is repeated p times, the link weight w_{ij} is updated as

$$w_{ij} = \begin{cases} w_{init} & ; \text{ if } w_{ij} \text{ does not exist} \\ w_{max} & ; \text{ else if } w_{ij} > w_{max} \\ w_{ij} + \delta & ; \text{ otherwise.} \end{cases} \qquad (4.2)$$

3) If the activation of the kernel K_i unit does not occur during a certain period p_1, the kernel unit K_i and all the link weights connected to the kernel unit $\mathbf{w}_i(= [w_{i1}, w_{i2}, \ldots])$ are removed from the SOKM (thus, representing the *extinction* of a kernel).

where $\xi_{ij}, w_{init}, w_{max}$, and δ are all positive constants. In 2) above, after the weight update, the excitation counters for both K_i and K_j, i.e. ε_i and ε_j, may be reset to 0, where appropriate. Then, both conditions 1) and 2) in the algorithm above also moderately agree with the rephrasing of Hebb's principle (Stent, 1973; Changeux and Danchin, 1976):

1. If two neurons on either side of a synapse are activated asynchronously, then that synapse is selectively weakened or eliminated[2].
2. If two neurons on either side of a synapse (connection) are activated simultaneously (i.e. synchronously), then the strength of that synapse is selectively increased.

4.2.2 Introduction of Decay Factors

Note that, to meet the second rephrasing above, a decaying factor is introduced within the link weight update algorithm (in Condition 1), to simulate the synaptic elimination (or decay). In the SOKM context, the second rephrasing is extended and interpreted such that i) the decay can always occur in time (though the amount of such decay is relatively small in a (very) short period of time) and ii) the synaptic decay can also be caused when the other kernel(s) is/are activated via the transmission of the activation of the kernel. In terms of the link weight decay within the SOKM, the former is represented by the factor ξ_{ij}, whereas the latter is under the assumption that the potential of the other end may be (slightly) lower than the one.

At the neuro-anatomical level, it is known that a similar situation to this occurs, due to the changes in the transmission rate of the spikes (Hebb, 1949; Gazzaniga et al., 2002) or the decay represented by e.g. long-term depression

---

[2]To realise the kernel unit connections representing the directional flows as described in Sect. 3.3.4, this rephrasing may slightly be violated.

(LTD) (Dudek and Bear, 1992). These can lead to modification of the above rephrasing and the following conjecture can also be drawn:

Conjecture 3: When kernel K_i is excited by input $\mathbf{x}$ and K_i also has connection to kernel K_j via the link weight w_{ij}, the activation of K_j is computed by the relation

$$K_j = \gamma w_{ij} K_i(\mathbf{x}) \qquad (4.3)$$

or

$$K_j = \gamma w_{ij} I_i \qquad (4.4)$$

where γ $(0 << \gamma \leq 1)$ is the decay factor, and I_i is defined as an indicator function

$$I_i = \begin{cases} 1 \text{ ; if the kernel } K_i(\mathbf{x}) \text{ is excited (i.e. when } K_i(\mathbf{x}) \geq \theta_K \\ 0 \text{ ; otherwise.} \end{cases} (4.5)$$

In the above, the indicator function I_i is sufficient to imitate the situation where an impulsive spike (or action potential) generated from one neuron is transmitted to the other via the synaptic connection (for a thorough discussion, see e.g. Gazzaniga et al., 2002), due to the excitation of the kernel K_i in the context of modelling the SOKM. The above also indicates that, apart from the regular input vector $\mathbf{x}$, the kernel can be excited by the secondary input, i.e. the transfer of the activations from other nodes, unlike conventional neural architectures. Thus, this principle can be exploited further for multi-domain data processing (in Sect. 3.3.1) by SOKMs, where the kernel can be excited by the transfer of the activations from other kernels so connected, without having such regular inputs.

In addition, note that another decay factor γ is introduced. This decay factor can then be exploited to represent a loss during the transmission.

4.2.3 Updating Link Weights Between (Regular) Kernel Units and Symbolic Nodes

In Figs. 3.4, 3.5, 3.7, and 3.8, various topological representations in terms of kernel memory have been described. Within these representations, the final network output kernel units are newly defined and used, in addition to regular kernel units, and it has been described that these output kernel units can be defined in various manners as in (3.16), (3.17), (3.18), (3.25), (3.28), or (3.30), without directly affecting the contents of the memory within each kernel unit. Such output units can thus be regarded as symbolic nodes (as in conventional connectionist models) representing the intermediary/internal states of the kernel memory and, in practice, exploited for various purposes, e.g. to obtain the pattern classification result(s) in a series of cognitive tasks (for a further discussion, see also Sects. 4.6 and 7.2).

Then, within the context of SOKM, the link weights between normal kernel units and such symbolic nodes as those representing the final network outputs can be either fixed or updated by [**The Link Weight Update Algorithm**] given earlier, depending upon the applications. In such situations, it will be sufficient to define the evaluation of the activation from such symbolic nodes in a similar manner to that in (3.12).

Thus, it is also said that the conventional PNN/GRNN architecture can be subsumed and evolved within the context of SOKM.

4.2.4 Construction/Testing Phase of the SOKM

Consequently, both the construction of an SOKM (or the training phase) and the manner of testing the SOKM are summarised as follows:

[Summary of Constructing A Self-Organising Kernel Memory]

Step 1)
- Initially ($cnt = 1$), there is only a single kernel in the SOKM, with the template vector identical to the first input vector presented, namely, $\mathbf{t}_1 = \mathbf{x}(1)$ (or, for the Gaussian kernel, $\mathbf{c}_1 = \mathbf{x}(1)$).
- If a Gaussian kernel is chosen, a unique setting of the radius σ may be used and determined *a priori* (Hoya, 2003a).

Step 2)
For $cnt = 2$ to {num. of input data to be presented}, do the following:

    Step 2.1)
- Calculate all the activations of the kernels K_i ($\forall i$) in the SOKM by the input data $\mathbf{x}(cnt)$, (e.g. for the Gaussian case, it is given as (3.8)).
- Then, if $K_i(\mathbf{x}(cnt)) \geq \theta_K$ (as in (3.12)), the kernel K_i is excited.
- Check the excitation of kernels via the link weights $\mathbf{w}_i$, by following the principle in **Conjecture 3**.
- Mark all the excited kernels.

    Step 2.2)
If there is no kernel excited by the input vector $\mathbf{x}(cnt)$, add a new kernel into the SOKM by setting its template vector to $\mathbf{x}(cnt)$.

> **Step 2.3)**
> Update all the link weights by following [**The Link Weight Update Algorithm**] given above.

In Step 1 above, initially there is no link weight but a single kernel in the SOKM and, later in Step 2.3, a new link weight may be formed, where appropriate.

Note also that Step 2.2 above can implicitly prevent us from generating an exponentially growing number of kernels, which is not taken into consideration by the original PNN/GRNN approaches. In another respect, the above construction algorithm can be seen as the extension/generalisation of the resource-allocating (or constructive) network (Platt, 1991), in the sense that 1) the SOKM can be formed to deal with multi-domain data simultaneously (in Sect. 3.3.1), which can potentially lead to more versatile applications, and 2) lateral connections are also allowed between the nodes within the sub-SOKMs responsible for the respective domains.

[Summary of Testing the Self-Organising Kernel Memory]

Step 1)
- Present input data $\mathbf{x}$ to the SOKM, and compute all the kernel activations (e.g. for the Gaussian case, this is given by (3.8)) within the SOKM.
- Check also the activations via the link weights $\mathbf{w}_i$, by following the principle in the aforementioned **Conjecture 3**.
- Mark all the excited kernels.

Step 2)
- Obtain the maximally activated kernel K_{max} (for instance, this is defined in (3.17)) amongst all the marked kernels within the SOKM.
- Then, if performing a classification task is the objective, the classification result can be obtained by simply restoring the class label η_{max} from the auxiliary memory attached to the corresponding kernel (or, by checking the activation of the kernel unit indicating the class label, in terms of the alternative kernel unit representation in Fig. 3.2).

As in the above, it is also said that the testing phase of the SOKM can take a similar step to constructing a Parzen window (Parzen, 1962; Duda et al., 2001)[3].

4.3 The Celebrated XOR Problem (Revisited)

In Sect. 2.3.2, the XOR problem as a benchmark test for general pattern classifiers has been solved in terms of a PNN/GRNN. Here, to see how an SOKM is actually constructed, we here firstly consider solving the same problem by means of an SOKM, as a straightforward pattern classification task.

Now, as in Sect. 2.3.2, let us consider the case where 1) Gaussian kernels with the unique radius setting of $\sigma = 1.0$ are chosen for the SOKM (with the ordinary kernel unit representation as in Fig. 3.1), 2) the activation threshold $\theta_K = 0.7$, and 3) the four input vectors to the SOKM consist of a pair of elements, i.e. $\mathbf{x}(1) = [0.1, 0.1]^T$, $\mathbf{x}(2) = [0.1, 1.0]^T$, $\mathbf{x}(3) = [1.0, 0.1]^T$, and $\mathbf{x}(4) = [1.0, 1.0]^T$. Then, by following the mechanism [**Summary of Constructing A Self-Organising Kernel Memory**] given earlier, the SOKM capable of classifying the four XOR patterns is constructed[4]:

[**Constructing an SOKM for Solving the XOR Problem**]

Step 1) (cnt=1:)
Initialise $\sigma = 1.0$ and $\theta_K = 0.7$.
Then, fix the centroid (template) vector of the first kernel K_1:

$\mathbf{c}_1 = \mathbf{x}(1) = [0.1, 0.1]^T$ and the class label $\eta_1 = 0$.
Step 2)
cnt=2:

Present $\mathbf{x}(2)$ to the SOKM (up to now, there is only a single kernel K_1 within the SOKM).

$K_1 = \exp(-\|\mathbf{x}(2) - \mathbf{c}_1\|_2^2/\sigma^2) = 0.4449$.

Thus, since $K_1(\mathbf{x}(2)) < \theta_K$, add a new kernel K_2 with setting $\mathbf{c}_2 = \mathbf{x}(2)$ and the class label $\eta_2 = 1$.

---

[3]However, to give a theoretical account for the multi-modal data processing aspect of SOKMs is beyond the scope of this book and thus must be an open issue, since the conventional approaches are mostly based upon a single domain pattern space (or hyper-plane); it does not seem to be sufficient to consider a simple extension of the single domain data representation to multiple domain situations, since in general the data points in the respective planes can be strongly correlated with each other.

[4]Needless to say, this is based upon a "one-shot" training scheme, as in PNNs/GRNNs.

---

cnt=3:

$$K_1 = \exp(-\|\mathbf{x}(3) - \mathbf{c}_1\|_2^2/\sigma^2) = 0.4449 \; (< \theta_K) \;,$$
$$K_2 = \exp(-\|\mathbf{x}(3) - \mathbf{c}_2\|_2^2/\sigma^2) = 0.1979 \; (< \theta_K) \;.$$

Thus, since there is no kernel excited by the input $\mathbf{x}(3)$, add a new kernel K_3, with $\mathbf{c}_3 = \mathbf{x}(3)$ and $\eta_3 = 1$.

cnt=4:

$$K_1 = \exp(-\|\mathbf{x}(4) - \mathbf{c}_1\|_2^2/\sigma^2) = 0.1979 \; (< \theta_K) \;,$$
$$K_2 = \exp(-\|\mathbf{x}(4) - \mathbf{c}_2\|_2^2/\sigma^2) = 0.4449 \; (< \theta_K) \;,$$
$$K_3 = \exp(-\|\mathbf{x}(4) - \mathbf{c}_3\|_2^2/\sigma^2) = 0.4449 \; (< \theta_K) \;.$$

Thus, again, since there is no kernel excited by $\mathbf{x}(4)$, add a new kernel K_4 with $\mathbf{c}_4 = \mathbf{x}(4)$ and $\eta_4 = 0$.

(Terminated.)

Then, it is straightforward that the above four input patterns can be correctly classified by following the procedure in [**Summary of Testing the Self-Organising Kernel Memory**] given earlier.

In the above, on first examination, constructing the SOKM takes similar steps for a PNN/GRNN, since there are four identical Gaussian kernels (or, RBFs) in a single network structure, as described in Sect. 2.3.2, and by regarding η_i ($i = 1, 2, 3, 4$) as the target values. (Therefore, it is also said that PNNs/GRNNs are subclasses of the SOKM.)

However, consider the situation where another set of input data, which, again, represent the XOR patterns, i.e. $\mathbf{x}(5) = [0.2, 0.2]^T$, $\mathbf{x}(6) = [0.2, 0.8]^T$, $\mathbf{x}(7) = [0.8, 0.2]$, and $\mathbf{x}(8) = [0.8, 0.8]^T$, is subsequently presented, during the construction of the SOKM. Then, despite all these patterns also being stored in general training schemes of PNNs/GRNNs, such redundant addition of kernels does not occur during the SOKM construction phase; these four patterns excite only the respective nearest kernels (due to the criterion (3.12)), all of which nevertheless yield the correct pattern classification results, and thus there are no further additional kernels. (In other words, this excitation evaluating process is viewed as testing of the SOKM.)

Therefore, from this observation, it is considered that by exploiting the local memory representation the SOKM acts as a pattern classifier which can simultaneously perform data pruning (or clustering), with proper parameter settings. In the next couple of simulation examples, the issue of the actual parameter setting for the SOKM is discussed further.

4.4 Simulation Example 1 – Single-Domain Pattern Classification

For the XOR problem, it has been discussed that the SOKM can be easily constructed to perform efficiently pattern classification of the XOR patterns. However, in that case, there were no link weights formed between the kernels.

In order to see how the SOKM is self-organised in a more realistic situation and how the activation via the link weights affects the performance of the SOKM, we then consider an ordinary single-domain pattern classification problem, namely, performing pattern classification tasks using several single-domain data sets, all of which are extracted from public databases.

For the choice of the kernel function in the SOKMs, a widely-used Gaussian kernel given in the form (3.8) is considered in the next two simulation examples, without loss of generality. Moreover, to simplify the problem for the purpose of tracking the behaviour of the SOKM, the third condition in [**The Link Weight Update Algorithm**] given in Sect. 4.2.1 (i.e. the kernel unit removal) is not considered in the simulation examples.

4.4.1 Parameter Settings

In the simulation examples, the three different domain datasets extracted from the original SFS (Huckvale, 1996), OptDigit, and PenDigit databases of "UCI Machine Learning Repository" at the University of California, were used as in Sect. 2.3.5. Thus, this yields three independent datasets for performing the classification tasks. The description of the datasets is summarised in Table 4.1. For the SFS dataset, the same encoding procedure as that in Sect. 2.3.5 was applied in advance to obtain the pattern vectors for the classification tasks.

Table 4.1. Data sets used for the simulation examples

Data Set	Length of Each Pattern Vector	Total Num. of Patterns in the Training Set	Total Num. of Patterns in the Testing Sets	Num. of Classes
SFS	256	540	360	10
OptDigit	64	1200	400	10
PenDigit	16	1200	400	10

Then, the parameters were arbitrarily chosen as summarised in Table 4.2 (in the left part). (As in Table 4.2, the combination of the parameters was chosen as uniquely as possible for all the three datasets, in order to perform the simulations in a similar condition.) During the construction phase of the SOKM, the settings $\sigma_i = \sigma$ ($\forall i$) and $\theta_K = 0.7$ were used for evaluating the excitation in (3.12). In addition, without loss of generality, the excitation of the kernels via the link weights was restricted only to the nearest neighbours (i.e. 1-nn) in the simulation examples.

Table 4.2. Parameters chosen for the simulation examples

Parameter	Data Set			
	For Single-Domain Pattern Classification			For Dual-Domain Pattern Classification (SFS+PenDigit)
	SFS	OptDigit	PenDigit	
Decaying Factor for Excitation γ	0.95	0.95	0.95	0.95
Unique Radius for Gaussian Kernel σ	8.0	5.0	2.0	8.0 (SFS) 2.0 (PenDigit)
Link Weight Adjustment Constant δ	0.02	0.02	0.02	0.02
Synaptic Decaying Factor $\xi_{i,j}$ $(\forall i, j)$	0.001	0.001	0.1	0.001
Threshold Value for Establishing Link Weights p	5	5	5	5
Initializing Value for Link Weights w_{init}	0.7	0.7	0.6	0.75
Maximum Value for Link Weights w_{max}	1.0	1.0	0.9	1.0

4.4.2 Simulation Results

Figures 4.1 and 4.2 show respectively the variations in the monotonically growing number of the kernels and link weights formed within the SOKM during
the construction phase. To check the relative growing numbers for the three
different domain datasets, a normalised scale of the pattern presentation number is used (in the x-axis). In the figures, each number $x(i)$ $(i = 1, 2, \ldots, 10)$
in the x-axis thus corresponds to the relative number of the pattern presentation, i.e. $x(i) = i \times$ {the total number of patterns in the training set}/10.

From the observation in Figs. 4.1 and 4.2, it can be said that the data
structure of the PenDigit dataset is relatively simple, compared to the other
two, since the number of kernels so generated is always the smallest, whereas
that of link weights is the largest. On the other hand, this is naturally considered by the evidence that, since the length of each pattern vector (i.e. 16)
as in Table 4.1 is the shortest amongst the three, the pattern space can be
constructed with a smaller number of data points in the PenDigit dataset
than the other datasets.

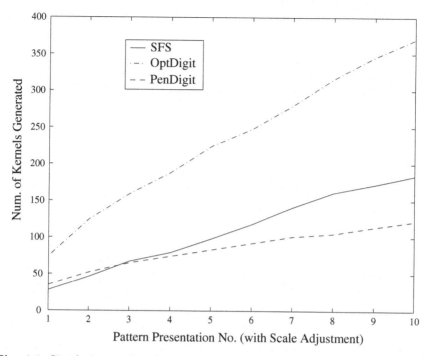

Fig. 4.1. Simulation results of single-domain pattern classification tasks – number of kernels generated during the construction phase of SOKM

4.4.3 Impact of the Selection σ Upon the Performance

It has been empirically confirmed that, as for the PNNs/GRNNs (Hoya and Chambers, 2001a; Hoya, 2003a, 2004b), a unique setting of the radii value within the SOKM gives a reasonable trade-off between the generalisation performance and the computational complexity. (Thus, during the construction phase of the SOKM, as described in Sect. 4.2.4, the parameter setting $\sigma_i = \sigma$ ($\forall i$) was chosen.)

However, as in PNNs/GRNNs, the selection of the radii σ_i still yields a significant impact upon the generalisation capability of SOKMs, amongst all the parameters. To investigate this further, the value σ is varied from the minimum Euclidean distance, calculated between all the pairs of pattern vectors in the training data set, to the maximum. For the three datasets, SFS, Opt-Digit, and PenDigit, both the maximum and minimum values so computed are tabulated in Table 4.3.

As in Figs. 4.3 and 4.4, the number of kernels generated as well as the overall generalisation capability of the SOKM is dramatically varied, according to the value σ; when σ is close to the minimum distance, the number of kernels is almost the same as the number of patterns in the dataset. In other words, almost all the training data are exhausted during the construction of

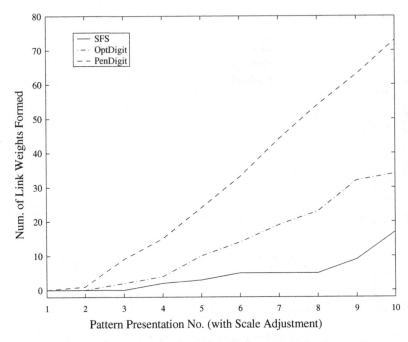

Fig. 4.2. Simulation results of single-domain pattern classification tasks – number of links formed during the construction phase of SOKM

Table 4.3. Minimum and maximum Euclidean distances computed amongst a pair of all the pattern vectors in the datasets

	Minimum Euclidean Distance	Maximum Euclidean Distance
SFS	2.4	11.4
OptDigit	1.0	9.3
PenDigit	0.1	5.7

the SOKM for such cases, which is computationally expensive. However, both Figs. 4.3 and 4.4 indicate that the decrease in the number of kernels does *not* always correspond to the relative degradation in terms of the generalisation performance. This tendency can also be confirmed by examining the number of correctly connected link weights (i.e. the number of link weights which establish connections between the kernels with identical class labels) as in Fig. 4.5:

Comparing Fig. 4.5 with Fig. 4.4, we observe that, for each data set, as the number of correctly connected link weights starts decreasing from the peak, the generalisation performance (as in Fig. 4.4) degrades sharply. From this observation, it can be justified that the values σ for the respective datasets in Table 4.2 were reasonably chosen. It can also be confirmed that with these

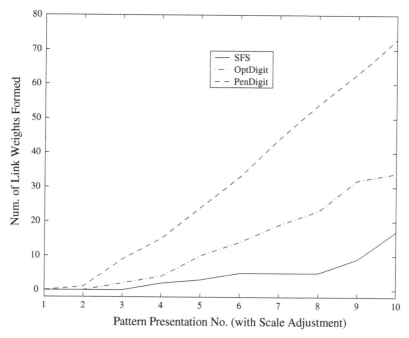

Fig. 4.3. Simulation results of single-domain pattern classification tasks – variations in the number of kernels generated with varying σ

values the ratio of the correctly connected link weights generated versus the wrong ones can be sufficiently high (i.e. the actual ratios were 2.1 and 7.3 for the SFS and OptDigit datasets, respectively, whereas the number of wrong link weights was zero for the PenDigit case).

4.4.4 Generalisation Capability of SOKM

Table 4.4 summarises the performance comparison between the SOKM so constructed (i.e. the SOKM of which all the pattern presentations for the construction is finished) using the parameters given in Table 4.2 and a PNN with the centroids found by the well-known MacQueen's k-means clustering algorithm. Then, the numbers of RBFs in the PNN responsible for the respective classes were fixed to those of the kernels within the SOKM.

As shown in Table 4.4, for the three datasets the overall generalisation performance of the SOKM is almost the same as/slightly better than the PNN + k-means approach, which verifies that the SOKM functions satisfactorily as a pattern classifier. However, it should be noted that, unlike ordinary clustering schemes, the number of kernels can be automatically determined by the unsupervised algorithm described in Sect. 4.2.1, and thus in this sense the manner of constructing the SOKM is more dynamic.

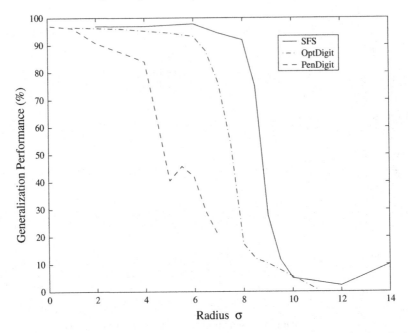

Fig. 4.4. Simulation results of single-domain pattern classification tasks – variations in the generalisation performance of the SOKM with varying σ

Table 4.4. Comparison of generalisation performance between the SOKM and a PNN using the k-means clustering algorithm

	Total Num. of Kernels Generated within SOKM	Generalisation Performance of SOKM	Generalisation Performance of PNN with k-means
SFS	184	91.9%	88.9%
OptDigit	370	94.5%	94.8%
PenDigit	122	90.8%	88.0%

4.4.5 Varying the Pattern Presentation Order

In the SOKM context, instead of the normal (or "well-balanced") pattern presentation (i.e. Pattern #1 of Digit /ZERO/, #1 of Digit /ONE/, ..., #1 of /NINE/, then Pattern #2 of Digit /ZERO/, #2 of Digit /ONE/, ..., etc), the manner of which is typical for constructing pattern classifiers, the order of pattern presentation can be varied 1) randomly or 2) as that for accommodating new classes (Hoya, 2003a) (i.e. Pattern #1 of Digit /ZERO/, #2 of Digit /ZERO/, ..., the last pattern of Digit /ZERO/, then Pattern #1 of Digit /ONE/, #2 of Digit /ONE/ ..., etc), since the construction is pattern-based. However, it has been empirically confirmed that these alternations do not affect either the number of kernels/link weights generated or the generalisation

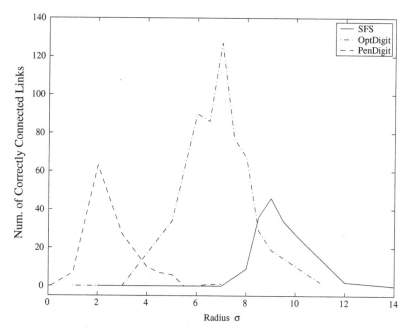

Fig. 4.5. Simulation results of single-domain pattern classification tasks – variations in the number of correctly connected links with varying σ

capability (Hoya, 2004a). This indicates that the self-organising architecture not only has the capability of accommodating new classes as PNNs (Hoya, 2003a) but also is robust to the varying conditions.

4.5 Simulation Example 2 – Simultaneous Dual-Domain Pattern Classification

In the previous example, it has been described that, within the context of pattern classification tasks, the SOKM yields a similar/slightly better generalisation performance, in comparison with a PNN/GRNN. However, it only reveals one of the potential benefits of the SOKM concept.

Here, we consider another practical example of multi-domain pattern classification task, in order to investigate further the behaviour of the SOKM, namely, a simultaneous dual-domain pattern classification in terms of the SOKM, which has not been considered in the conventional neural network studies, as stated earlier.

In the simulation example, an integrated SOKM consisting of two sub-SOKMs is designed to imitate the situation where a specific voice sound input to a particular area (i.e. the area responsible for auditory modality) of memory excites not only the auditory area but in parallel or *simultaneously* the visual (thus the term *"simultaneous dual-domain pattern classification"*), on the ground that the appropriate built-in feature extraction mechanisms for the respective modalities are provided within the system. This is thus somewhat relevant to the issues of modelling the "associations" between different cognitive modalities, or, in a more general context, the "concept formation" (Hebb, 1949; Wilson and Keil, 1999) or mental imagery, in which several perceptual processes are concurrent and, in due course, united together (i.e. "data-fusion"), in which the integrated notion or, what is called, *Gestalt* (see Section 9.2.2) formation occurs.

4.5.1 Parameter Settings

Then, for the actual simulation, we consider the case using both the SFS (for digit voice recognition) and PenDigit (for digit character recognition) datasets (Hoya, 2004a), each of which constitutes a sub-SOKM responsible for the corresponding specific domain data, and the cross-domain link weights (or, the associative links) between a certain number of kernels within both the sub-SOKMs are formed by the link weight algorithm given in Sect. 4.2.1. (Then, an artificial data-fusion of both the datasets is thereby considered.) The parameters for updating the link weights to perform the dual-domain task are summarised in the last column of Table 4.2. For the formation of the associative links between the two sub-SOKMs, the same values as those for the ordinary links (i.e. the link weights within the sub-SOKM) given in Table 4.2 were chosen (except the synaptic decay factor $\xi_{ij} = \xi = 0.0005$ $(\forall i, j)$).

In addition, for modelling such a cross-modality situation, it is natural to consider that the order of presentation may also affect the formation of the associative links. However, without loss of generality, the patterns were presented alternatively across the two training data sets (viz., the pattern vector SFS #1, PenDigit #1, SFS #2, PenDigit #2, ...) in the simulation.

4.5.2 Simulation Results

In Table 4.5 (in both the second and fourth columns), the overall generalisation performance of the dual-domain pattern classification task is summarised. In the table, the item "Sub-SOKM(i) → Sub-SOKM(j)" (i.e. Sub-SOKM(1) indicates a single sub-SOKM responsible for the SFS data set, whereas Sub-SOKM(2) for the PenDigit) denotes the overall generalisation performance obtained by excitations of the kernels within Sub-SOKM(j), due to the transfer of the excitations in Sub-SOKM(i) via the associative links from the kernels within Sub-SOKM(i).

4.5.3 Presentation of the Class IDs to SOKM

In the three simulation examples given so far, the auxiliary parameter η_i to store the class ID was given whenever a new kernel is added in to the SOKM

Table 4.5. Generalisation performance of the dual-domain pattern classification task

| | Generalisation Performance (GP)/Num. Excited Kernels via the Associative Links (NEKAL) | | | |
| | Without Constraint | | With Constraints on Links | |
	GP	NEKAL	GP	NEKAL
SFS	86.7%	N/A	91.4%	N/A
PenDigit	89.3%	N/A	89.0%	N/A
Sub-SOKM(1) → (2)	62.4%	141	73.4%	109
Sub-SOKM(2) → (1)	88.0%	125	97.8%	93

and fixed to the same value as that of the current input data. However, unlike ordinary connectionist schemes, within the SOKM context it is not always necessary to set the parameter η_i at the same time as the input pattern is presented. Then, it is also possible to set η_i *asynchronously* where appropriate. In Chap. 7, this principle will be justified within a more general context of *"reinforcement learning"* (Turing, 1950; Minsky, 1954; Samuel, 1959; Mendel and McLaren, 1970).

Within this principle, we next consider a slight modification to the link weight updating algorithm, in which the class ID η_i is used to regulate the generation of the link weights, and show that such a modification can yield the performance improvement in terms of generalisation capability.

4.5.4 Constraints on Formation of the Link Weights

As described above, within the SOKM context, the class IDs can be given at any time, dependent upon applications. Then, we here consider the case where the information about the class IDs is known *a priori*, which is also not untypical in practice (though this modification may violate the strict sense of "unsupervised-ness"), and see how such a modification gives an impact upon the performance of the SOKM.

In this principle, the link weight update algorithm given in Sect. 4.2.1 is modified by taking the constraints on the link weights into account (the modified part is underlined below):

[The Modified Link Weight Update Algorithm]

1) if the link weight w_{ij} is already established, decrease the value according to:

$$w_{ij} = w_{ij} \times \exp(-\xi_{ij}) \tag{4.6}$$

---

2) If the subsequent excitation of a pair of kernels K_i and K_j $(i \neq j)$ occurs (the excitation is judged by (3.12)) and repeated for p times <u>and if the class IDs of both the kernels K_i and K_j are identical,</u> the link weight w_{ij} is updated as

$$w_{ij} = \begin{cases} w_{init} & ; \text{ if } w_{ij} \text{ does not exist} \\ w_{max} & ; \text{ else if } w_{ij} > w_{max} \\ w_{ij} + \delta & ; \text{ otherwise.} \end{cases} \qquad (4.7)$$

3) If the activation of the kernel K_i unit does not occur during a certain period p_1, the kernel unit K_i and all the link weights $\mathbf{w}_i$ are removed from the SOKM (representing the "*extinction*" of the kernel).

Simulation Results

With the modification above, the overall generalisation performance of the SOKM can be improved as in Table 4.5 (in the fourth column).

Moreover, Fig. 4.6 compares the number of links generated for both the cases of the SOKM grown by the link weight update algorithm with/without the constraints on the class IDs. As in the figure, for all the types of the link weights (i.e. SFS only, PenDigit, and the associative link weights in between the two datasets), it is observed that the number of links with the constraints above is smaller than that without them. This is considered simply because the "wrong" connections of the kernels (i.e. the links which connect the kernels with different class IDs) were avoided during the construction phase (Nevertheless, in a wider sense, this sort of constraints must be dealt within the general context of learning (to be described in Chap. 7).)

4.5.5 A Note on Autonomous Formation of a New Category

In Sect. 4.5.3, it has been described that the class IDs can be given at any time, and the actual setting of the parameter η_i (or making connections to the kernels indicating class IDs, by exploiting the modified kernel unit representation shown in Fig. 3.2) depends upon the application. If we reinterpret this description in terms of the modified kernel unit in Fig. 3.2, it is considered that the autonomous formation of new categories can occur during the construction phase of the SOKM, in terms of the following principle:

1) A new kernel unit is created within the SOKM. (At this point, there is no link weight(s) generated for this new kernel.)

2) At some point later, a new category is given, as a new kernel within the SOKM.

3) Then, the new kernel unit is connected to the kernel indicating the category by the link weight.

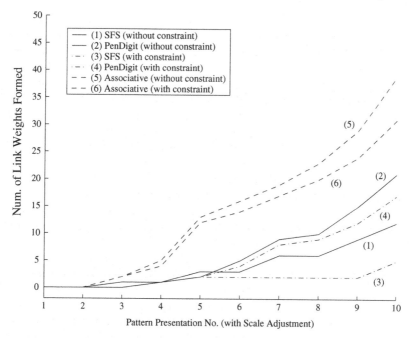

Fig. 4.6. Simulation results of dual-domain pattern classification tasks – number of links formed during the construction phase

In 3) above, it is considered that the new kernel given at 1) has already been connected to the kernel(s) which either does or does not indicate other categories/classes. However, it is also considered that, in terms of the link weight update algorithm given in Sect. 4.2.1, only the link weights (or those with maximum values) which survived during the construction phase eventually represent the actual categories/classes, whilst the remaining relatively weaker link weights are not effective enough to describe the categories/classes (or extinct from the SOKM). In this principle, it is thus evident that binding the kernels with too many classes/categories can be automatically avoided. We will turn back to the issue of category (or concept) formation in Chap. 9 (Sect. 9.2.2).

4.6 Some Considerations for the Kernel Memory in Terms of Cognitive/Neurophysiological Context

As described so far, the kernel memory concept is based upon a simple connection mechanism of multiple kernel units. The connection rule between the kernel units such as given in Sect. 4.2 for SOKMs is followed by the original neuropsychological principle of Hebbian learning (Hebb, 1949), in which when a kernel A is excited and one of the link weights is connected to kernel B, the excitation of kernel A is transferred to kernel B via the link weight (in **Conjecture 2** in Sect. 4.2). To date, Hebb's principle (Hebb, 1949) has still been influential in the areas not limited to computational but general neuroscience. (His speculations which appeared in (Hebb, 1949), are really remarkable, considering that the examination of real brain tissues was then very difficult.)

Followed by the neurophysiological findings of the existence of the so-called "hand-cells" within the inferior temporal cortex of macaque (Gross et al., 1972), Desimone et al. (Desimone et al., 1984) carefully examined the behaviour of these cells and reported that such cells selectively respond to the visual stimuli of hand images but not to other complex ones such as facial or comb-like images (cf. Hubel and Wiesel, 1977).

It is therefore natural to consider that the memory-based pattern recognition approach of the KM principle sufficiently matches the aforementioned neurophysiological findings; a single (or multiple) kernel unit(s) represents the cells that selectively respond to particular objects.

In the cognitive scientific context, such cells are quite often referred to as the so-called "gnostic units" (or grand-mother cells) to represent higher perceptual functions (Gazzaniga et al., 2002), which have appeared in the controversial issue of how the object perception is actually performed. In the concept of grand-mother cells, it is assumed that only a single cell placed on the top of hierarchical coding system is responsible for the perception of an object.

It has then been argued that the concept of grand-mother cells (or the hierarchical coding scheme) cannot explain the situation 1) if a gnostic unit dies, a sudden loss for the particular object is experienced, which is neither intuitively nor naturally considered to happen, and 2) how to perceive novel objects. In contrast to the grand-mother cell concept, the ensemble coding scheme (for a general description, see e.g. Gazzaniga et al., 2002) has also been considered amongst the cognitive science community, in which the activation of multiple (i.e. not single) higher-order neurons are involved in parallel in order to perceive an object. (This is hence related to the issue of *concept formation*. We will revisit this issue in Chap. 9 (Sect. 9.2.2).) In a recent study (Tsunoda et al., 2001), the neuroscientific finding which supports the principle of ensemble coding is reported.

Nevertheless, as described so far in both the present and previous chapters, it is considered that the KM concept can still suffice the aforementioned conditions required for both the hierarchical and ensemble coding schemes and be exploited to provide the models/practical examples. Note that, throughout this book, the KM concept is not treated as the basis for describing precisely various neuro-anatomical phenomena which occur within the real brain, as in the conventional artificial neural network principle (cf. Kohonen, 1997), but rather exploited for the (limited) utility in modelling behavioral/higher-order functions related to the mind.

4.7 Chapter Summary

In this chapter, the kernel memory concept described in the previous chapter has been exploited to develop a constructive network architecture, namely, the self-organising kernel memory (SOKM). The behaviours of SOKM have been discussed through some simulation examples given in the context of pattern classification tasks. In the simulation examples, the SOKMs have been compared with the existing connectionist models.

Then, in the description, it has been revealed that the SOKM exhibits the following seven main features:

- A single kernel unit can be ultimately regarded as the smallest memory element that simultaneously performs pattern classification (cf. the neuropsychological basis on RBFs made by Poggio and Edelman, 1990).
- The architecture of the kernel memory is intuitive and straightforward: The parameter tuning algorithm can be relatively simple, without suffering from numerical instability, unlike the conventional neural network architectures. Moreover, within the SOKM principle, the manner of construction (or self-organisation)/testing within the network can be fully traced, where required. In addition, there is no clear cut between the construction (or training) and testing phase of the SOKM.
- Flexible network configuration – straightforward and robust incremental training/network forgetting and accommodation of new classes (Hoya, 2003a), inherited from the properties of PNNs/GRNNs. Moreover, unlike conventional artificial neural network schemes, an instance (represented by a kernel unit) is allowed to belong simultaneously to multiple classes.
- Unlike the original PNN/GRNN approaches, the SOKM itself can exhibit capability in data pruning.
- There exist essentially no topological constraints within the KM concept (unlike conventional neural architectures, such as MLP-NNs or SOFMs). However, a number of useful fixed topological representations depending upon applications are also possible within a single learning principle, where appropriate, which has not been taken into account within the original PNNs/GRNN context.
- Related to the above, the SOKM can itself process multiple domain (i.e. "data-fusion") or temporal data, simultaneously/in parallel, both of which are considered to be significant for modelling the complex data processing as performed by real brain. In this respect, the SOKM can also be seen as the extension/generalisation to the resource-allocating network (Platt, 1991). However, these features, as well as the aforementioned flexible network configuration property, are not usually treated within the context of conventional artificial neural networks; even within the modern approaches as SVMs these aspects have been considered little, whilst a great number of theoretically related/performance improvement issues have been reported (see e.g. Vapnik, 1995; Hearst, 1998; Christianini and Taylor, 2000).

- By means of the kernel memory concept, the dynamic memory architecture (or self-evolutionary system) can be designed to provide both the distributed and local representation of memory, depending upon the application.

In the subsequent chapters, the concept of kernel memory will be given as a foundation for modelling various psychological functions which are postulated as the keys to constitute eventually the artificial mind system.

Part II

Artificial Mind System

5

The Artificial Mind System (AMS), Modules, and Their Interactions

5.1 Perspective

The previous two chapters have been devoted to establishing the novel artificial neural network concept, namely the kernel memory concept, for the foundation of the artificial mind system (AMS).

In this chapter, a global picture of the artificial mind system, which can be seen as a multi-input multi-output system, is presented. It is seen that the artificial system consists of a total of fourteen modules and their interactions, each of which plays a central role to model the corresponding cognitive/psychological function of the mind. The concept of modules to represent the respective functionalities in the AMS is originally motivated/inspired from psychological studies (Fodor, 1983; Hobson, 1999).

In the subsequent Chaps. 6–10, more general accounts of the respective modules (as those implemented within the two exemplar models) and their mutual interactions within the AMS, as well as the justifications from other studies, are given in detail.

Thus, the content of the present chapter (and the later chapters) often and essentially differs from those in the previous three chapters, in that the issues treated hereafter will be sometimes more macroscopic accounts of the artificial mind system, rather than only ending up with minor engineering justifications of the artificial neural substrate established in the previous three chapters, though the kernel memory concept described in the last two chapters remains important in the general model of the AMS.

In Chap. 10 (Sects. 10.6 and 10.7), a couple of models exploiting the several modules within the AMS will also be given, with a practical implementation to construct intelligent pattern classification systems.

Tetsuya Hoya: *Artificial Mind System – Kernel Memory Approach*, Studies in Computational Intelligence (SCI) **1**, 83–94 (2005)
www.springerlink.com

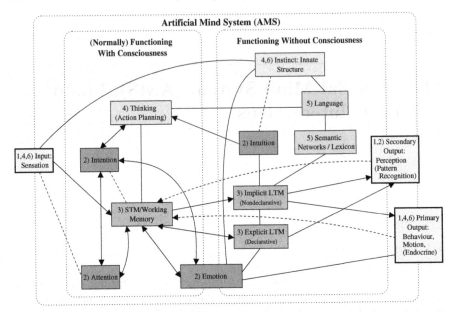

Fig. 5.1. A schematic diagram of the artificial mind system (AMS) – as a multi-input multi-output (MIMO) system consisting of 14 "modules"; one single input, two output modules, and the remaining 11 modules, each of which represents the corresponding cognitive/psychological function, and their mutual interactions

Table 5.1. The background studies to provide the accounts for the respective modules within the AMS shown in Fig. 5.1. Each number indicates the categories/main studies to provide the notions of the respective modules

---

1) Input/Outputs of the Artificial Mind System
2) Psychology & Cognitive Neuroscience
3) Memory (Connectionism & Psychology)
4) Artificial Intelligence, Signal Processing, Robotics (Mechanics),
   & Optimisation (Control Theory)
5) Linguistics (Language), Connectionism, & Optimisation
   (e.g. Graph Theory)
6) Innate Structure: Developmental Studies, Ecology, Genetics, etc.

5.2 The Artificial Mind System – A Global Picture

As shown in Fig. 5.1, the artificial mind system can be macroscopically regarded as a multi-input multi-output (MIMO) system, consisting of 14 modules, i.e. one single input, two outputs, and the remaining 11 modules representing the respective cognitive/psychological functions.

As in Table 5.2, it is considered that the four modules, attention, intention, STM/working memory, and thinking, normally function with consciousness, whilst the other six, i.e. instinct, intuition, language, both the explicit and

Table 5.2. Classification of the modules within the AMS in terms of the functionality with consciousness/without consciousness; it is considered that a total of five modules function with consciousness, whereas the seven operate without consciousness. The emotion module can have both consciousness and subconsciousness states

(Normally) Functioning with Consciousness	Functioning without Consciousness
(1) Attention	**(1)** Emotion
(2) Emotion	**(2)** Instinct
(3) Intention	**(3)** Intuition
(4) STM/Working Memory	**(4)** Language
(5) Thinking (Action Planning)	**(5)** Explicit LTM
	(6) Implicit LTM
	(7) Semantic Networks/Lexicon

implicit LTM, and semantic networks/lexicon, function without consciousness (or, subconsciously).

The LTM can be divided into two types of modules, i.e. the explicit and implicit (or the declarative and nondeclarative) LTM modules as in Fig. 5.1[1]. In contrast, the module "emotion" can be exceptionally regarded as a module functioning either with or without consciousness, depending upon situations. Moreover, as in Table 5.2, it is considered that the module "language" lies in/functions in parallel to the semantic networks/lexicon. From the LTM aspect, the language module also appears as a built-in (but still dynamic) structure (i.e. such as the learning mechanism of grammar) which is closely tied to the module representing semantic networks/lexicon (to be described in Chaps. 8 and 9).

The number(s) shown in each module indicates the corresponding relevant disciplines/categories (as shown in Table 5.1) in order to give the concrete accounts/notions for the functionalities, e.g. the functionality of the module "intention" takes into account of (at least) the principles within psychology.

As in Fig. 5.1, the "input" represents sensation and the output from the AMS can be classified into two types of "outputs"; i) the primary outputs which represent actual behaviour, endocrine, motion, or determine the direction, and ii) the secondary outputs obtained as a cause of perception. The perceptual activities in the latter generally involve pattern recognition of the internal feedbacks/external stimulus-oriented inputs arriving at the working/short-term memory (STM).

In Fig. 5.1, the modules are *connected* to the others via the links, representing the interactions (or, more appropriately, some form of information transmission) in between. As depicted in Fig. 5.1, there are three types of links denoted in 1) *solid lines* with and 2) without mono- and bi-directional

---

[1]Unless denoted otherwise, hereafter the mere "LTM" denotes both the explicit and implicit LTM modules.

Fig. 5.2. The kernel memory concept (in Chaps. 3 and 4) – especially, as the foundation of the memory-oriented modules within the AMS, i.e. both the explicit and implicit LTM, STM/working memory, and semantic networks/lexicon modules

arrows, and 3) *dashed lines*, which respectively indicate the modules involving the (mono-/bi-)directional information transmission, those functioning essentially in parallel, and the modules indirectly interrelated.

Then, as indicated in Fig. 5.2, to represent the memory modules within the AMS – the two types of LTM, STM, and semantic networks/lexicon – the kernel memory (KM) concept, which has been proposed as a new form of artificial neural network/connectionist model in Chaps. 3 and 4, plays a crucial role (to be discussed further in Chap. 8), though as described later, for the other modules such as emotion, input: sensation, intuition, and so forth, the KM concept also underlies.

The overall structure of the AMS in Fig. 5.1 is thus closely tied to the psychological concept in terms of *modularity of mind*, which is originally motivated/inspired from the psychological studies (Fodor, 1983; Hobson, 1999).

Then, it is seen that the modules within the AMS generally agree with the principle of Hobson (Hobson, 1999), i.e. the respective constituents for describing consciousness in Table 1.1 (on page 5), except that the constituent "orientation" can also be dealt within the framework of the intention module in the AMS context (to be described later in Chap. 10).

In addition, it is stressed that, since the stance for developing an artificial mind system in this book is based upon the speculation from the behaviour of human-beings/phenomena occurred in brain, it does not necessarily involve the controversial place-adjustment, within the neuroscientific context, between the regions in real brain and the respective psychological functions, in order to imitate and realise their functionalities by means of substances other than real brain tissue or cells.

5.2.1 Classification of the Modules Functioning With/Without Consciousness

As discussed earlier, the four modules in the AMS, i.e. attention, intention, STM/working memory, and thinking, normally function with consciousness, whilst the other six, i.e. instinct, intuition, both the explicit and implicit LTM, language, and semantic networks/lexicon, are considered to function without consciousness[2]. The remaining module, i.e. emotion, is the cross-over module between consciousness and subconsciousness.

In the AMS, it is intuitively considered that those functioning *consciously* are meant to be such modules that the functionalities, where necessary, can be (almost) fully controlled and their behaviours can be monitored in any detail (if required) by other consciously functioning module(s). However, this sometimes may be violated, depending upon situations (or, more specifically, the resultant data transmissions as the cause of the data processing within themselves/mutual interactions in between), i.e. some modules may well be considered to function with consciousness (though the judgement of consciousness/subconsciousness may often differ from one way of view to another[3]). In such irregular cases, some data can be easily lost from those functioning consciously or the leakage within the information transmission between the modules can occur in due course.

For instance, the emotion module functions with consciousness, when the attention mechanism is largely affected by the incoming inputs (arriving at the STM/working memory module), but the module can be affected subconsciously, depending upon the overall internal states of the AMS. In such a situation, the current environment/condition for the AMS can even be said to abnormal, e.g. the energy left is low, or, the temperature surrounding the robot is no longer tolerable (though this is not explicitly shown in Fig. 5.1).

In a real implementation, it could be helpful to attach the respective *consciousness/subconsciousness states* to the modules, the status of which can also be counted as the internal state within the AMS.

5.2.2 A Descriptive Example

Now, we consider a descriptive example to determine what kind of processing of the modules within the AMS is involved and how their mutual interactions occur for a specific task.

It is evident that one single example is not sufficient to explain fully how the AMS works in Fig. 5.1, however, in general, there can be countless numbers of scenarios to compose for validating the AMS completely, and it is virtually impossible to cover all the scenarios in the context. Hence, we limit ourselves

---

[2]As will be discussed later in Chap. 8, though the explicit LTM module itself is considered to work subconsciously, the access to the contents from the STM module is performed consciously.

[3]In the author's view, the terminology of *consciousness/subconsciousness* has been established from various psychological studies, which are largely based upon the interpretation/translation of the phenomena occurring in the brain by human-beings; ultimately speaking, no definitive manner has been found to determine whether it is functioning with or without consciousness, and thus, the judgement is not objective but rather subjective. In this book, we do not go further into the discussion of this issue.

to consider how we can interpret the following simple story in terms of the AMS:

> *"At the concert last night, I was listening to my favourite tune, Rachmaninoff's Piano Concerto No. 2, so as to let my hair down. But, I became a bit angry when my friend suddenly interrupted my listening by her whispering in my right ear and thus I immediately responded with a 'shush' to her ..."*

Q.) How do we interpret the above scenario in terms of the artificial mind system (AMS) shown in Fig. 5.1?

The answer to the above question can be described as follows:

A.) Overall, this can be interpreted in such a way that, by the sudden stimulus input (friend's voice sound), 1) the attention module was affected (this is then related to *selective attention*), 2) hence the emotional states of the AMS were suddenly varied, and, as a consequence, 3) vocalised the word "shush" to stop her whispering. More specifically, it is considered that the following four steps are involved:

Step 1) Prerequisite (initial formation)
Step 2) (Regular) incoming data processing
Step 3) Interruption of the processing in Step 2)
Step 4) Making real actions

Now, let us consider each of the steps above in more detail:

Step 1) Prerequisite (initial formation)

Step 1.1) Within the **LTM** (i.e. the episodic/semantic part of the memory) of the AMS, the tune of Rachmaninoff's Piano Concerto No. 2 has already been stored[4] so that the pattern recognition can be straightforwardly performed and the corresponding kernels can be excited by the (encoded) orchestral sound.

Step 1.2) Then, the subsequent pattern recognition result of each phrase that can be represented by a kernel unit (without loss of generality, provided that the whole tune can be divided into multiple phrases which have already been stored within the LTM) is

---

[4]In terms of the kernel memory, it is considered that the tune can be stored in the form of e.g. "a chain of kernel units", where each kernel unit represents some form of musical elementary unit (such as a phrase or note, etc) obtained by the associated feature extraction mechanism. Such chain can be constructed within the principles of kernel memory concept described in Chaps. 3 and 4. In a more general sense, the construction of such kernel-chains can be seen as the "learning" process (to be described at full length in Chap. 7).

considered as a series of the **secondary (or perceptual) output(s)** of the AMS (as in Fig. 5.1), which will also be subsequently fed back to the **STM/working memory** and eventually control the **emotional states**.

Step 1.3) The module emotion consists of some (i.e. a multiple number of) potentiometers (*four*, say, to describe 1) pleasure, 2) anger, 3) grief, and 4) joy). The corresponding kernel units representing the respective phrases are synaptically connected to the first & fourth potentiometers (i.e., the potentiometers representing pleasure and joy, through the learning process). Thus, if the subsequent excitation of such kernel units is a result of the external stimuli (i.e. by listening to the orchestral playing), the excitation can also be transferred to the potentiometers and in due course cause the changes in the potentials.

Step 1.4) Moreover, as indicated in Fig. 5.1, the values of the emotional states are directly transferred to/connected with the **primary outputs** (to cause real actions, such as resting the arms, smiling on the face, or other parts of the body, endocrine, and so forth).

Step 1.5) In addition, the **input: sensation** module may involve preprocessing; specifically, such as sound activity detection (SAD), feature extraction, where appropriate, or blind signal/source separation (BSS) (see e.g. Cichocki and Amari, 2002) mechanisms. In Sect. 8.5, an example of such preprocessing mechanisms, i.e. a combined neural memory, which exploits PNNs, and blind signal processing (BSP) for extracting the specified speech signal from the mixture of simultaneously uttered voice sounds is given.

Step 2) (Regular) incoming data processing

Just before the friend's voice arrives at the **input module (sensation)**, the incoming input is processed (with first priority) within the **STM/working memory**, which is the sound (or the feature data) coming from the orchestra, due to the **attention** module. Then, this had maintained the two out of four potentials (representing pleasure and joy) being positive (and relatively higher compared to the rest) within the module **emotion**.

Therefore, a total of seven modules in the AMS (i.e. in the descriptions above, the contexts related to the corresponding seven modules are denoted in *bold*) and their mutual interactions are considered to be involved for Steps 1) and 2) as in the below:

Modules involved in Steps 1-2)

1) Attention   5) Primary Outputs
2) Emotion   6) Secondary Output
3) Input: Sensation   7) STM/Working Memory
4) LTM (Explicit/Implicit)

Mutual interactions occurring in Steps 1-2)

- **Input: Sensation $\longrightarrow$ STM/Working Memory**:
  Arrival of the orchestral sound.
- **STM/Working Memory $\longrightarrow$ LTM**:
  Accessing the episodic/semantic or declarative memory of the orchestral sound.
- **Implicit/Explicit LTM $\longrightarrow$ Secondary (Perceptual) Output**:
  Perception/pattern recognition of the orchestral sound.
- **Secondary (Perceptual) Output $\longrightarrow$ STM/Working Memory**:
  The feedback input (where appropriate); the pattern recognition results of the orchestral sound.
- **STM/Working Memory $\longrightarrow$ Attention**:
  Arrival of the orchestral sound.
- **Attention $\longrightarrow$ STM/Working Memory $\longrightarrow$ Emotion**:
  Maintaining the current emotional states due to the subsequent orchestral sound inputs.
- **Emotion – Primary Outputs (Endocrine)**
- **Emotion $\longrightarrow$ STM/Working Memory $\longrightarrow$ Implicit LTM $\longrightarrow$ Primary Outputs (Motions)**:
  Making real actions, such as resting the arms, endocrine, etc.

Step 3) Interruption of the processing in Step 2)

When the friend's whispering arrived at the **STM/working memory**, with a relatively higher volume/duration sufficient to affect the **attention** module (or, as in the prerequisite in Step 1) above, the feedback inputs to the STM/working memory, after the (subsequent) *perception* of her voice), the **emotional states** were greatly affected. This is since, in such a situation, the friend's voice varied the selective attention, which could no longer maintain the current positive potentials within the two emotional states, thereby causing the drop in these values, and eventually the value of the second potentiometer (anger) may have become positive.

Modules involved in Step 3)

    1) Attention   4) LTM (Explicit/Implicit)
    2) Emotion   5) Secondary output
    3) Input: Sensation 6) STM/Working Memory

Mutual interactions occurring in Step 3)

- **Input: Sensation $\longrightarrow$ STM/Working Memory**:
  Arrival of the friend's whispering sound.
- **STM/Working Memory $\longrightarrow$ LTM** and
- **Implicit/Explicit LTM $\longrightarrow$ Secondary (Perceptual) Output**:
  Perception, pattern recognition of the friend's voice.
- **Secondary (Perceptual) Output $\longrightarrow$ STM/Working Memory**:
  The feedback input; the pattern recognition results from the friend's voice.
- **STM/Working Memory $\longrightarrow$ Attention**:
  Effect upon the selective attentional activity due to the arrival of the friend's voice.
- **Attention $\longrightarrow$ STM/Working Memory $\longrightarrow$ Emotion**:
  Varying the current emotional states as the cause of the sudden friend's voice.

As in the above, it is considered that a total of six modules are involved and mutually interacted for Step 3). In the above, albeit denoted explicitly, the sixth data flow **attention $\longrightarrow$ STM/working memory $\longrightarrow$ emotion** also indicates a possible situation that the emotional states are varied due to the intention module as a cause of the thinking process performed via the thinking module, since the thinking module is considered to function in parallel with the STM/working memory. In such a case, the emotional states are varied e.g. *after* some semantic analysis of her voice and its access to the declarative (or explicit) LTM, representing the *reasoning* process of the interruption.

Step 4) Making real actions

    Step 4.1) In many situations, it is considered that, as aforementioned, Step 3) above also involves the process within the **thinking** module (functioning in parallel with the **STM/working memory**), regardless of its consciousness state.

    Step 4.2) Then, the AMS performed the decision-making to issue the command to "increase" the value of the second **emotional**

state (anger) via the STM/working memory and eventually vo-
calise the sound "shush" to her, due to the episodic content of
memory (acquired by *learning* or *experience*) e.g. that represents
the general notions, *"whilst music playing, one has to be quiet till
the end/interval"* and *"to stop one's talking, making the sound
"shush" is often effective"* (this is under the condition that the
word can be understood (in English), i.e. the module **language** is
involved), the context of which can also be interpreted by the ref-
erences to the **LTM** or the **semantic networks/lexicon** (both
of which are considered to function in parallel).

Step 4.3) The action of vocalising the word involves the processes
(mainly) within the STM/working memory invoked by the **inten-
tional** activity ("to make the sound") and the **primary output**.

Step 4.4) Moreover, provided that the action of vocalising is (recognised as) effective (due to both the thinking and **perception** modules), i.e. to successfully stop her whispering, this indicates that
the action taken (due to the accesses to the implicit LTM) had
been successful to resume the previous emotional states (represented by the emotion module, i.e. the two relatively higher potentials representing "pleasure" and "joy" than the other two, with
paying **attention** to the incoming orchestral sound).

Modules involved in Step 4)

1) Attention	6) Primary Outputs
2) Emotion	7) Secondary Output
3) Intention	8) Semantic Networks/Lexicon
4) Language	9) STM/Working Memory
5) LTM (Explicit/Implicit)	10) Thinking

Mutual interactions occurring in Step 4)

- **STM/Working Memory – Thinking Module**:
  These two modules are normally functioning in parallel,
  for the decision-making process to deal with the sudden
  changes in the emotional states.
- **STM/Working Memory ⟶ LTM or Semantic
  Networks/Lexicon**:
  Accessing the verbal sound "shush", the **language** mod-
  ule is also involved to recognise the word in English.

- **Intention and STM/Working Memory** $\longrightarrow$ **Implicit LTM** $\longrightarrow$ **Primary Outputs**:
  Vocalising the word "shush".
- **STM/Working memory** $\longrightarrow$ **LTM** $\longrightarrow$ **Secondary (Perceptual) Output**:
  Perception/pattern recognition of the friend's responses.
- **Secondary (Perceptual) Output** $\longrightarrow$ **STM/Working Memory**:
  The feedback input (where appropriate); the pattern recognition results of the friend's stopping her whispering.
- **STM/Working Memory and Thinking** $\longrightarrow$ **Implicit LTM (Procedural Memory)**:
  The processing was invoked after the perception that the vocalising "shush" was effective via the pattern recognition results of her responses.
- **Implicit LTM – Emotion**:
  Varying the emotional states which represent the previous states.
- **Emotion** $\longrightarrow$ **STM/Working Memory** $\longrightarrow$ **Attention**:
  Maintaining the current emotional states by paying again attention to the orchestral sound.

For Step 4), a total of ten modules and their mutual interactions are therefore considered to be involved, as in the above.

As in the scenario example examined above, it is evident that a total of 12 modules (indicated in *boldfaces*) and their mutual interactions, which constitutes most of the AMS in Fig. 5.1, are involved even within this simple scenario.

The four subsequent Chaps. 6–10 are then devoted to the detailed descriptions of the modules within the AMS. The detailed accounts of the two unattended modules in this example, instinct and intuition, are thus left to the later Chaps. 8 and 10 (i.e. in Sects. 8.4.6 and 10.5, respectively).

Moreover, a concrete model for pattern classification tasks, which exploits the four modules representing attention, intuition, LTM, and STM, and the extended model will appear in Chap. 10.

5.3 Chapter Summary

This chapter have firstly provided a global picture of the artificial mind system. The AMS has been shown to consist of a total of 14 modules, each of which is responsible for specific cognitive/psychological function, and involves their mutual interactions. The modular approach is originally inspired/motivated from the psychological studies in Fodor (1983); Hobson (1999). Then, the behaviour of the AMS and how the associated modules interact with each other have been analysed by examining a simple scenario.

It has also been proposed that the kernel memory concept established in the last three chapters plays a key role, especially for consolidating the memory mechanisms within the AMS.

In the five succeeding Chaps. 6–10, the discussion is moved to the more detailed accounts of the respective modules and their mutual interactions.

Sensation and Perception Modules

6.1 Perspective

In any kind of creature, both the mechanisms of *sensation* and *perception* are indispensable for continuous living, e.g. to find edible plants/fruits in the forest, or to protect themselves from attack by approaching enemies. To fulfill these aims, there are considered to be two different kinds of information processes occurring in the brain: 1) extraction of useful features amongst the flood of information coming from the sensory organs equipped and 2) perception of the current surroundings based upon the features so detected in 1) for planning the next actions to be taken. Namely, the sensation mechanism is responsible for the former, whereas the latter is the role of the perception mechanism.

In this chapter, we highlight the two modules within the AMS, i.e. the **sensation** and **perception** modules within the sensory inputs area. In the AMS, it is considered that the sensation module receives information from the outside world and then converts it into the data which can be efficiently handled within the AMS, whilst the perception module plays a central role to represent what is currently occurring in the AMS and generally yields the pattern recognition results by accesses to the memory modules, which can be used for further data processing.

It is considered that the sensation module can consist of multiple pre-processing units. As aforementioned, one of the important aspects of the sensation module is how to detect useful information in noisy situations. More specifically, this topic is related to *noise reduction* in the signal processing field. In this context, we will consider a practical example of noise reduction based totally upon a signal processing application, namely the reduction of noise in stereophonic speech signals, in which the binaural data processing of humans is modelled and evaluated through extensive simulation examples in Sect. 6.2.2. As will be described later, the functionality of the perception module is closely related to the memory modules in Chap. 8. In Sect. 8.5, we will also consider another example relevant to noise reduction, i.e. speech

Tetsuya Hoya: *Artificial Mind System – Kernel Memory Approach*, Studies in Computational Intelligence (SCI) **1**, 95–116 (2005)
www.springerlink.com

extraction in cocktail party situations, which exploits both the concept of pre-processing units within the sensation module and memory modules.

It will also be described further in the next chapter (Chap. 7) that the functionality, as well as the formation, of both the two modules, sensation and perception, are closely interrelated with each other via the concept of general learning.

6.2 Sensory Inputs (Sensation)

As described in the previous chapter, the artificial mind system shown in Fig. 5.1 (on page 84) can be macroscopically viewed as an input-output system. In the figure, the module **sensory inputs: sensation** functions as the receiver for the sensory input data arriving at the AMS. Then, the role of the sensation module is also to pre-process/encode the raw data received into the feature data (where appropriate) that can be efficiently handled with the other modules within the AMS.

As in Fig. 5.1, the data processed within the sensation module are all fed forward to the **STM/working memory** module.

In general, it is considered that humans are inherently equipped with five sensors to interact with the outside world, i.e. sensors for visual, auditory, gustatory (taste), olfactory, and tactile input[1].

For developing artificial intelligence or real robots, such sensors as those detecting e.g. infra-red, radioactivity, or other specific rays, depending upon situations, can be considered (as those alternative to visual sensory inputs), in addition to the aforementioned five sensors. Note that, within the AMS in Fig. 5.1, though the number of sensory inputs arriving at the AMS may be varied, it is considered that it does not essentially affect the overall layout of the modules within the AMS[2].

Within the AMS, it is assumed that the input data received by the sensation module are either raw sensory or pre-coded data (or the data obtained via a certain process of feature extraction). Then, the sensation module is

---

[1]However, it is said that the role of an actual sensory organ of humans is not always restricted to acquire only a single sensory mode but rather to process multi-modal data in parallel (i.e. data-fusion). For instance, the biological mechanism of the human ears exploits the tactile information which is received as the sound pressure by ear drum, converted into electrical activities, and eventually transferred via the auditory nerve to the auditory cortex (for more details, see e.g. Gazzaniga et al., 2002), or the tongue can sense not only the taste but simultaneously the weight or temperature of objects. Moreover, it is considered that many of the sensory organs also function as actuators.

[2]However, we should bear in our mind that sensor combinations different from those of humans (i.e. other than the five sensors) could completely vary the structure within the respective modules of the AMS, as the cause of the learning/evolution process. This issue is then related to the so-called "Mind-Body" problem.

Raw Sensory
Input Data

Input Data
for Modules in AMS

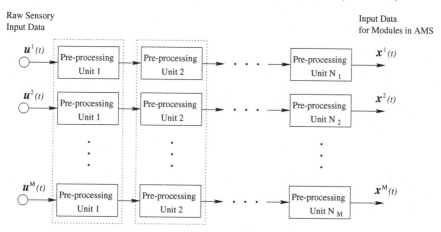

Fig. 6.1. An illustrative diagram of the sensory inputs (sensation) module – defined as a cascade of pre-processing units. Note that the boxes in dotted line indicate the necessity (i.e. in signal processing wise) of the utility of multi-sensory input data, rather than single, for some particular pre-processing

also responsible for converting the raw sensory into pre-coded data by means of feature detecting mechanisms, where appropriate, in order to reduce the redundancy and process efficiently within the modules of the AMS.

6.2.1 The Sensation Module – Given as a Cascade of Pre-processing Units

As illustrated in Fig. 6.1, it is considered that the sensation module is composed of several submodules, each representing a specific pre-processing mechanism.

In Fig. 6.1, $\mathbf{u}^i(t)$ $(i = 1, 2, \ldots, \mathbf{M})$ denotes the i-th raw sensory input data measurement to the sensation module arriving at time instance t and $\mathbf{x}^i(t)$ are the corresponding feature data signals obtained after a series of pre-processing stages. In Fig. 6.1, the i-th sensation module can be (approximately) represented in a cascading form of N_i pre-processing submodules (or units), each of which transforms the raw data into other useful representation, where appropriate.

For instance, for the processing of auditory signals, such pre-processing as source localisation/direction of arrival (DOA) estimation (see e.g. Hudson, 1981), sound activity detection (SAD), noise reduction (NR) (see e.g. Davis, 2002), or (blind) signal extraction (BSE)/separation (BSS) (see e.g. Cichocki and Amari, 2002), all of which are active areas of study in signal processing, may be involved. (In Fig. 6.1, note that the boxes in dotted line indicate the necessity (in signal processing wise) of the utility of multi-sensory input data, rather than single, for some particular pre-processing.)

In the cognitive scientific context, it is generally considered that the cochlea of a human ear plays a central role to pre-process the auditory information in a similar fashion to spatio-temporal coding mechanism (Barros et al., 2000; Rutkowski et al., 2000; Barros et al., 2002), whilst for the visual information both the retinal and the V1-V4 areas of the brain contribute to the feature extraction (see e.g. Gazzaniga et al., 2002), which can also be in a wider sense regarded as a spatio-temporal coding mechanism. In recent studies, the spatio-temporal scheme has also been exploited for olfactory recognition tasks (White et al., 1998; Hoshino et al, 1998; Lysetskiy et al., 2002).

Then, it appears interesting, since the spatio-temporal scheme (i.e., represented by a subband structure) can be ultimately considered as one of the universal pre-processing mechanisms for the sensory information acquired.

In the next section 6.2.2, a signal processing based example of stereophonic noise reduction in speech signals is described. Moreover, later in Sect. 8.5, an example showing how to exploit a combined blind signal extraction technique with the aid of neural memory (thus related to the **memory** modules) for the extraction of speech signals in cocktail party situations will also be given.

In addition, although the actual pre-processing mechanisms e.g. BSE (Cichocki and Amari, 2002), NR, or SAD can be realised by means of exploiting the existing signal processing/pattern recognition techniques, the principle of hierarchy similar to Neocognitron developed by Fukushima (Fukushima, 1975) may be exploited, and thereby a more biologically plausible neural-based model (Brian et al., 2001) could be devised. We will revisit this issue in Chap. 7.2.

6.2.2 An Example of Pre-processing Mechanism – Noise Reduction for Stereophonic Speech Signals (Hoya et al., 2003b; Hoya et al., 2005, 2004c)

Here, we consider a practical example of the pre-processing mechanism based upon a signal processing application – noise reduction for stereophonic speech signals by a combined cascaded subspace analysis and adaptive signal enhancement (ASE) approach (Hoya et al., 2003b; Hoya et al., 2005). The subspace analysis (see e.g. Oja, 1983) is a well-known approach for various estimation problems, whilst adaptive signal enhancement has long been a topic of great interest in the adaptive signal processing area of study (see e.g. Haykin, 1996).

In this example, a multi-stage sliding subspace projection (M-SSP) is firstly used, which operates as a sliding-windowed subspace noise reduction processor, in order to extract the source signals for the post-processors, i.e. a bank of adaptive signal enhancers. Thus, the role of the M-SSP is to extract the (monaural) source signal. In each stage of M-SSP, a subspace decomposition algorithm such as eigenvalue decomposition (EVD) can be employed.

Then, for the actual signal enhancement, a bank of modified adaptive signal (line) enhancers is used. For each channel, the enhanced signal obtained

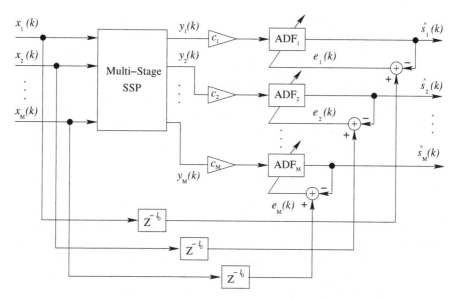

Fig. 6.2. Block diagram of the proposed multichannel noise reduction system (Hoya et al., 2003b; Hoya et al., 2005, 2004c) – a combined multi-stage sliding subspace projection (M-SSP) and adaptive signal enhancement (ASE) approach; the role of M-SSP is to reduce the amount of noise on a stage-by-stage basis, whereas the adaptive filters (denoted ADF_i) compensate for the spatio-temporal information at the respective channels, e.g. in two-channel situations (i.e. $M = 2$), to recover the stereophonic image

from the M-SSP is given to the adaptive filter as the source signal for the compensation of the stereophonic image. The principle of this approach is that the quality of the outputs of the M-SSP will be improved by the adaptive filters (ADFs).

In the general case of an array of sensors, the M-channel observed sensor signals $x_i(k)$ $(i = 1, 2, ..., M)$ can be represented by

$$x_i(k) = s_i(k) + n_i(k), \ (i = 1, 2, \ldots, M) \tag{6.1}$$

where $s_i(k)$ and $n_i(k)$ are respectively the target and noise components within the observations $x_i(k)$.

Figure 6.2 illustrates the block diagram of the proposed multichannel noise reduction system, where $y_i(k)$ denotes the i-th signal obtained from the M-SSP, and $\hat{s}_i(k)$ is the i-th enhanced version of the target signal $s_i(k)$.

Here, we assume that the target signals $s_i(k)$ are speech signals arriving at the respective sensors, that the noise process is zero-mean, additive, and uncorrelated with the speech signals, and that $M = 2$. Thus, under the assumption that $s_i(k)$ are all generated from one single speaker, it can be considered that the speech signals $s_i(k)$ are strongly correlated with each other and thus that we can exploit the property of the strong correlation for noise reduction by a subspace method.

In other words, we can reduce the additive noise by projecting the observed signal onto the subspace of which the energy of the signal is mostly concentrated. The problem here, however, is that, since speech signals are usually non-stationary processes, the correlation matrix can be time-variant. Moreover, it is considered that the subspace projection reduces the dimensionality of the signal space, e.g. a stereophonic signal pair can be reduced to a monaural signal.

Noise Reduction by Subspace Analysis

The subspace projection of a given signal data matrix contains information about the signal energy, the noise level, and the number of sources. By using a subspace projection, it is thus possible to divide approximately the observed noisy data into the subspaces of the signal of interest and the noise (Sadasivan et al., 1996; Cichocki et al., 2001; Cichocki and Amari, 2002).

Let $\mathbf{X}$ be the available data in the form of an $L \times M$ matrix

$$\mathbf{X} = [\mathbf{x}_1, \mathbf{x}_2, \ldots, \mathbf{x}_M] , \tag{6.2}$$

where the column vector $\mathbf{x}_i$ $(i = 1, 2, \ldots, M)$ is written as

$$\mathbf{x}_i = [x_i(0), x_i(1), \ldots, x_i(L-1)]^T \ (T: \text{transpose}) . \tag{6.3}$$

Then, the EVD of the autocorrelation matrix of $\mathbf{X}$ (for $M < L$) is given by

$$\mathbf{X}^T\mathbf{X} = \mathbf{V}\mathbf{\Sigma}\mathbf{V}^T, \tag{6.4}$$

where the matrix $\mathbf{V} = [\mathbf{v}_1, \mathbf{v}_2, \ldots, \mathbf{v}_M] \in \mathfrak{R}^{M \times M}$ is orthogonal such that $\mathbf{V}^T\mathbf{V} = \mathbf{I}_M$ and $\mathbf{\Sigma} = diag(\sigma_1, \sigma_2, \ldots, \sigma_M) \in \mathfrak{R}^{M \times M}$, with eigenvalues $\sigma_1 \geq \sigma_2 \geq \ldots \geq \sigma_M \geq 0$. The columns in $\mathbf{V}$ are the eigenvectors of $\mathbf{X}^T\mathbf{X}$. The eigenvalues in $\mathbf{\Sigma}$ contain some information about the number of signals, signal energy, and the noise level. It is well known that if the signal-to-noise ratio (SNR) is sufficiently high (see e.g. Kobayashi and Kuriki, 1999), the eigenvalues can be ordered in such a manner as

$$\sigma_1 > \sigma_2 > \cdots > \sigma_s \gg \sigma_{s+1} > \sigma_{s+2} \cdots > \sigma_M \tag{6.5}$$

and the autocorrelation matrix $\mathbf{X}^T\mathbf{X}$ can be decomposed as

$$\mathbf{X}^T\mathbf{X} = [\mathbf{V}_s \ \mathbf{V}_n] \begin{bmatrix} \mathbf{\Sigma}_s & \mathbf{O} \\ \mathbf{O} & \mathbf{\Sigma}_n \end{bmatrix} [\mathbf{V}_s \ \mathbf{V}_n]^T , \tag{6.6}$$

where $\mathbf{\Sigma}_s$ contains the s largest eigenvalues associated with s signals with the highest energy (i.e., $\sigma_1, \sigma_2, \ldots, \sigma_s$) and $\mathbf{\Sigma}_n$ contains $(M - s)$ eigenvalues $(\sigma_{s+1}, \sigma_{s+2}, \ldots, \sigma_M)$. It is then considered that $\mathbf{V_s}$ contains s eigenvectors

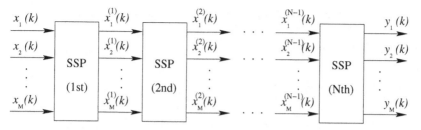

Fig. 6.3. Block diagram of the multi-stage SSP (up to the N-th stage) using M-channel observations $x_i(k)$ $(i = 1, 2, \ldots, M)$; for noise reduction, it is considered the amount of noise after the j-th SSP is smaller than that after the $j - 1$-th SSP operation

associated with the signal part, whereas $\mathbf{V}_n$ contains $(M - s)$ eigenvectors associated with the noise. The subspace spanned by the columns of $\mathbf{V}_s$ is thus referred to as the signal subspace, whereas that spanned by the columns of $\mathbf{V}_n$ corresponds to the noise subspace.

Then, the signal and noise subspaces are mutually orthogonal, and orthonormally projecting the observed noisy data onto the signal subspace leads to noise reduction. The data matrix after the noise reduction $\mathbf{Y} = [\mathbf{y}_1, \mathbf{y}_2, \ldots, \mathbf{y}_M]$, where $\mathbf{y}_i = [y_i(0), y_i(1), \ldots, y_i(L - 1)]^T$, is given by

$$\mathbf{Y} = \mathbf{X}\mathbf{V}_s\mathbf{V}_s^T \tag{6.7}$$

which describes the orthonormal projection onto the signal space.

This approach is quite beneficial to practical situations, since we do not need to assume/know in advance the locations of the noise sources. For instance, in stereophonic situations, since both the speech components $\mathbf{s}_1$ and $\mathbf{s}_2$ are strongly correlated with each other, even if the rank is reduced to one for the noise reduction purpose (i.e., by taking only the eigenvector corresponding to the eigenvalue with the highest energy σ_1), it is still possible to recover $\mathbf{s}_i$ from $\mathbf{y}_i$ by using adaptive filters (denoted ADF_i in Fig. 6.2) as the post-processors.

The Sliding Subspace Projection

In many applications, the subspace projection above is employed in a batch mode. Here, we instead consider on-line batch algorithms for adaptively estimating the subspaces which are operated in a cascade form.

Figure 6.3 shows a block diagram for the N-stage SSP. As in the figure, the observed signals $x_i(k)$ are processed through multiple stages of SSP.

The concept of the multi-stage structure was motivated from the work of Douglas and Cichocki (Douglas and Cichocki, 1997), in which natural gradient type algorithms (Cichocki and Amari, 2002) are used in a cascading form for blind decorrelation/source separation.

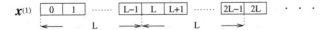

Conventional frame–based subspace analysis

Multi–stage sliding subspace projection operation

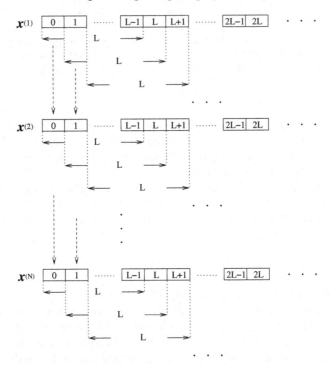

Fig. 6.4. Illustration of the multi-stage SSP operation (with the data-reusing scheme in (6.8)); as on the top, in conventional subspace approaches, the analysis window (or frame) is always distinct, whereas an overlapping window (of length L) is introduced at each stage for the M-SSP

Within the scheme, note that since the SSP acts as a sliding-window noise reduction block and thus that M-SSP can be viewed as an N-cascaded version of the block. To illustrate the difference between the M-SSP and the conventional frame-based operation (e.g. Sadasivan et al., 1996), Fig. 6.4 is given. In the figure, $\mathbf{x}^{(j)}$ denotes a sequence of the M-channel output vectors from the j-th stage SSP operation, i.e., $\mathbf{x}^{(j)}(0), \mathbf{x}^{(j)}(1), \mathbf{x}^{(j)}(2), \ldots (j = 1, 2, \ldots, N)$, where $\mathbf{x}^{(j)}(k) = [x_1^{(j)}(k), x_2^{(j)}(k), \ldots, x_M^{(j)}(k)]$ $(k = 0, 1, 2, \ldots)$. As in the figure, the SSP operation is applied to a small fraction of data (i.e. the sequence of L samples) using the original input at time instance k in each stage and outputs only the signal counterpart for the next stage. This operation is repeated at the subsequent time instances $k + 1, k + 2, \ldots$, and thus the name "sliding".

The Multi-Stage SSP

Then, given the previous L past samples for each channel at time instance k ($\geq L$) and using (6.7), the input matrix to the j-th stage SSP $\mathbf{X}^{(j)}(k)$ ($L \times M$) can be given:

1) The Scheme With Data-Reusing (Hoya et al., 2003b; Hoya et al., 2005)

$$\mathbf{X}^{(j)}(k) = \begin{bmatrix} \mathbf{P}\mathbf{X}^{(j)}(k-1)\mathbf{V}_s^{(j)}(k-1)\mathbf{V}_s^{(j)}(k-1)^T \\ \mathbf{x}^{(j-1)}(k) \end{bmatrix} ,$$

$$\mathbf{P} = [\mathbf{0}_{(L-1) \times 1}; \mathbf{I}_{L-1}] \ (L-1 \times L) \tag{6.8}$$

2) The Scheme Without Data-Reusing (Hoya et al., 2004c)

$$\mathbf{X}^{(j)}(k) = \mathbf{X}^{(j-1)}(k)\mathbf{V}_s^{(j-1)}(k)\mathbf{V}_s^{(j-1)}(k)^T \tag{6.9}$$

where $\mathbf{V}_s^{(j)}$ denotes the signal subspace matrix obtained at the j-th stage and

$$\mathbf{x}^{(0)}(k) = \mathbf{x}(k),$$

$$\mathbf{X}^{(j)}(0) = \begin{bmatrix} \mathbf{0}_{(L-1) \times M} \\ \mathbf{x}^{(j-1)}(0) \end{bmatrix}.$$

In (6.8) (i.e. the operation with the data-reusing scheme), note that, in contrast to (6.9), the first $(L-1)$ rows of $\mathbf{X}^{(j)}(k)$ are obtained from the previous SSP operation in the same (i.e. the j-th) stage, whereas the last row is taken from the data obtained from the original observation ($j = 0$)/the data obtained in the previous (i.e. the $(j-1)$-th) stage. Then, at this point, as in Fig. 6.4, the new data contained in the last row vector $\mathbf{x}^{(j-1)}(k)$ (i.e. the data from the previous stage) always remains intact, whereas the first $(L-1)$ row vectors, i.e. those obtained by the product $\mathbf{P}\mathbf{X}^{(j)}(k-1)\mathbf{V}_s^{(j)}(k-1)\mathbf{V}_s^{(j)}(k-1)^T$ will be replaced by the subsequent subspace projection operations. It is thus considered that this recursive operation is similar to the concept of data-reusing (Apolinario et al., 1997) or fixed point iteration (Forsyth et al., 1999) in which the input data at the same data point is repeatedly used for improving the convergence rate in adaptive algorithms.

Then, the first row of the new input matrix $\mathbf{X}^{(j)}(k)$ given in (6.8) or (6.9) corresponds to the M-channel signals after the j-th stage SSP operation $\mathbf{x}^{(j)}(k) = [x_1^{(j)}(k), x_2^{(j)}(k), \ldots, x_M^{(j)}(k)]^T$:

$$\mathbf{x}^{(j)}(k) = \mathbf{X}^{(j)}(k)^T \mathbf{q} ,$$

$$\mathbf{q} = [1, 0, 0, \ldots, 0]^T \ (L \times 1) . \tag{6.10}$$

Thus, the output from the N-th stage SSP $\mathbf{y}(k) = [y_1(k), y_2(k), \ldots, y_M(k)]^T$ yields:

$$\mathbf{y}(k) = \mathbf{x}^{(N)}(k) \ . \tag{6.11}$$

In (6.8) or (6.9), since the input data used for the j-th stage SSP are different from those at the $j-1$-th stage, it is expected that the subspace spanned by $\mathbf{V}_s$ can contain less noise than that obtained at the previous stage.

In addition, we can intuitively justify the effectiveness of using M-SSP as follows: for large noise variance and very limited numbers of samples (this choice must, of course, relate to the stationarity of the noise), a single stage SSP may perform only rough or approximate decomposition to both the signal and noise subspace. In other words, we are not able to ideally decompose the noisy sensor vector space into a signal subspace and its noise counterpart with a single stage SSP. In the single stage, we rather perform decomposition into a signal-plus-noise subspace and a noise subspace (Ephraim and Trees, 1995). For this reason, applying M-SSP gradually reduces the noise level. Eventually, the outputs obtained after the N-th stage SSP, $y_i(k)$, are considered to be less noisy than the respective inputs $x_i(k)$ and sufficient to be used for the input signal to the signal enhancers.

As described, the orthonormal projection of each observation $x_i(k)$ onto the estimated signal subspace by the M-SSP leads to reduction of the noise in each channel. However, since the projection is essentially performed using only a single orthonormal vector which corresponds to the speech source, this may cause the distortion of the stereophonic image in the extracted signals $y_1(k)$ and $y_2(k)$. In other words, the M-SSP is performed only to recover the single speech source from the two observations $x_i(k)$.

Related to the subspace-based noise reduction as a sliding window operation, it has been shown that a truncated singular value decomposition (SVD) operation is identical to an array of analysis-synthesis finite impulse response (FIR) filter pairs connected in parallel (Hansen and Jensen, 1998). It is then expected that this approach still works when the number of the sensors M is small, as in ordinary stereophonic situations (i.e. $M = 2$).

Two-Channel Adaptive Signal Enhancement

Without loss of generality, we here consider a two-channel adaptive signal enhancer (ASE, or alternatively, dual adaptive signal enhancer, DASE) in order to compensate for the stereophonic image from the extracted signals $y_1(k)$ and $y_2(k)$ by M-SSP.

As in Fig. 6.2, since the observations $x_i(k)$ are true stereophonic signals (albeit noisy), it is considered that applying adaptive signal enhancers to the extracted signals by M-SSP can lead to the recovery of the stereophonic image in $\hat{s}_i(k)$ by exploiting the stereophonic information contained in the error signals $e_i(k)$, since the extracted signal counterparts are strongly correlated with the corresponding signal of interest. The adaptive filters then function to adjust both the delay and amplitude of the signal in the respective channels.

Note that, in Fig. 6.2, the delay elements are inserted to delay the reference signals $x_i(k)$ by half the length of the adaptive filters L_f:

$$l_0 = \frac{L_f - 1}{2}. \tag{6.12}$$

This is to shift the centre lag of the reference signals to the centre of the adaptive filters, i.e. to allow not only the positive but also negative direction of time by the adaptive filters.

This scheme is then somewhat related to direction of arrival (DOA) estimation using adaptive filters (Ko and Siddharth, 1999) and similar to ordinary adaptive line enhancers (ALEs) (see e.g. Haykin, 1996). However, unlike a conventional ALE, the reference signal in each channel is not taken from the original input but the observation $x_i(k)$. Moreover, in the context of stereophonic noise reduction, the role of the adaptive filters is considered to be deviated from the original DOA, as described above.

In addition, in Fig. 6.2, c_i are arbitrarily chosen constants and used to adjust the scaling of the corresponding input signals to the adaptive filters. These scaling factors are normally necessary, since the choice will affect the initial tracking ability of the adaptive algorithms in terms of stereophonic compensation and may be determined *a priori* with keeping a good-trade off between the initial tracking performance and the signal distortion. Finally, as in Fig. 6.2, the enhanced signals $\hat{s}_i(k)$ are obtained simply from the respective filter outputs, where for the two channel case $\hat{s}_i$ $(i = 1, 2)$ represent the signals after the stereophonic noise reduction.

6.2.3 Simulation Examples

Here, we consider some simulation examples with the following observations representing a stereophonic environment:

$$x_1(k) = a \times s_1(k) + n_1(k),$$
$$x_2(k) = a \times s_2(k) + n_2(k), \tag{6.13}$$

where $s_1(k)$ and $s_2(k)$ correspond respectively to the left and right channel speech signal arriving at the respective microphones, $n_1(k)$ and $n_2(k)$ are the noise components, and the constant "a" controls the input SNR. In stereophonic situations, the two channel speech components $s_1(k)$ and $s_2(k)$ are strongly correlated with each other and approximated by:

$$s_1(k) = \mathbf{h}_1^T(k)\mathbf{s}(k),$$
$$s_2(k) = \mathbf{h}_2^T(k)\mathbf{s}(k), \tag{6.14}$$

where $\mathbf{h}_i(k) = [h_i(0), h_i(1), \ldots, h_i(L_s-1)]^T$ $(i = 1, 2)$ are the impulse response vectors of the acoustic transfer functions between the signal (speech) source and the microphones with length L_s, and $\mathbf{s}(k) = [s(k), s(k-1), \ldots, s(k-L_s+1)]^T$ is the speech source signal vector.

Therefore, it is considered that the respective stereophonic speech components $s_i(k)$ ($i = 1, 2$) are generated from one speech source using two (sufficiently long) filters $\mathbf{h}_i$ and, in reality, the stereophonic speech components $s_i(k)$ are strongly correlated with each other.

Then, the objective here is to eliminate both the noise components $n_1(k)$ and $n_2(k)$ from the corresponding observation $x_1(k)$ and $x_2(k)$.

For the simulation examples given here, the length of the analysis matrix was fixed to 32 for an SSP, whilst, for the DASE, the standard normalised least mean square (NLMS) algorithm (see e.g. Haykin, 1996) was used to adjust the adaptive filter coefficients. For each adaptive filter, the learning constant was chosen as 0.5, and the filter lengths were 51, the latter selection of which neither precedence effect (or, alternatively, the Haas effect) nor echo effect was not considered to occur (Hugonnet and Walder, 1998). Moreover, the scalar constants c_i were empirically adjusted to 0.1, which moderately suppressed the distortion and satisfied a good trade-off between a reasonable stereophonic image compensation and signal distortion.

Figure 6.5 (a) shows the left and right channel signals of a real speech of the sentence "Pleasant zoos are rarely reached by efficient transportation" uttered (in English) by a male speaker, (b) the noisy speech (assuming the input SNR=3(dB)), (c) the enhanced speech obtained from nonlinear spectral subtraction (NSS) algorithm (Martin, 1994; Xie and Van Compernolle, 1996; Gustafsson et al., 1999; Martin, 2001; Gustafsson et al., 2003), which is one of the classical and most commonly used methods for (single-channel) noise reduction, applied independently to each channel, (d) the enhanced speech by a single stage SSP (with the data-reusing scheme given in (6.8) (Hoya et al., 2003b; Hoya et al., 2005)), and (e) the enhanced speech by the combined SSP and DASE method, respectively.

For the simulation example shown in Fig. 6.5, the noise components $n_1(k)$ and $n_2(k)$ are assumed to be two i.i.d. random variables generated from the Normal distribution.

In Fig. 6.6, another example is shown, using real speech from the same sentence as in Fig. 6.5 but uttered by a different male speaker. In contrast to the results in Fig. 6.5, the number of stages for the M-SSP was increased to eight and the SSP without data-reusing scheme given in (6.9) was applied for the results in Fig. 6.6. In Fig. 6.6 (e), it is shown that the amount of the noise components is greatly reduced, in comparison with Fig. 6.6 (e). This indicates the effectiveness of applying the multi-stage SSP.

As shown in Figs. 6.5 (d,e) and 6.6 (e,f), the overall shape of the speech buried in noise is mostly recovered, whilst some voiced parts are eliminated or greatly changed in shape in the enhanced speech by NSS in Figs. 6.5 and 6.6 (c). This can be confirmed by listening; the enhanced speech obtained from the NSS sounds "hollow" besides the annoying additive musical tonal noise. In contrast, the enhanced speech by the methods based upon SSP and DASE does not have such artifacts or distortion. It can also be confirmed that

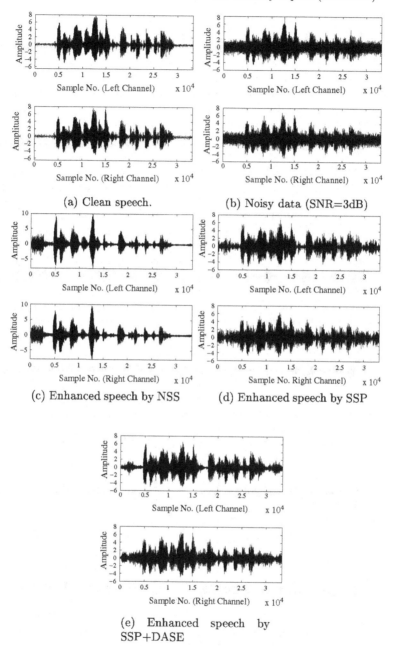

(a) Clean speech.

(b) Noisy data (SNR=3dB)

(c) Enhanced speech by NSS

(d) Enhanced speech by SSP

(e) Enhanced speech by SSP+DASE

Fig. 6.5. A simulation example of stereophonic noise reduction – using a real stereo-phonic speech for the sentence (in English) "Pleasant zoos are rarely reached by efficient transportation" uttered by a male speaker – using the data-reusing scheme in (6.8)

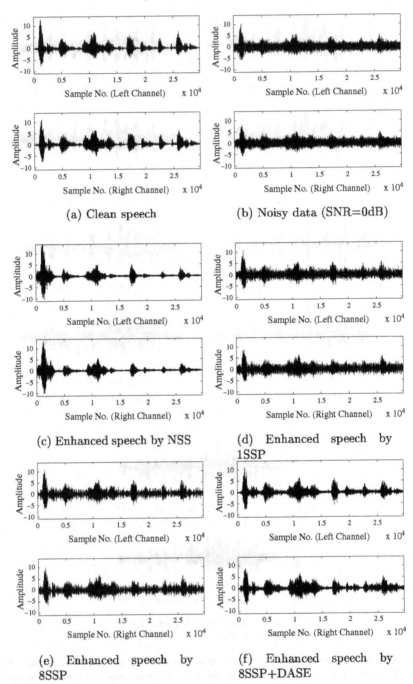

(a) Clean speech

(b) Noisy data (SNR=0dB)

(c) Enhanced speech by NSS

(d) Enhanced speech by 1SSP

(e) Enhanced speech by 8SSP

(f) Enhanced speech by 8SSP+DASE

Fig. 6.6. Another simulation example – using the speech data of the same sentence as Fig. 6.5 uttered by a different male speaker – using (6.9), i.e. the scheme without the data-reusing, instead of (6.8)

the two-channel signals obtained from the M-SSP sound rather dual-mono, as described earlier, or that the spatial image is gone, whilst the stereophonic image is recovered by the post-processing of DASE.

Next, to further confirm the aforementioned effectiveness of the combined noise reduction scheme, we consider two objective measurements, the segmental gain[3] and cepstral distance, both of which are commonly used measurements for evaluating the quality in speech enhancement (see e.g. Deller et al., 1993; Le Bouquin-Jennes et al., 1997).

Then, the segmental gain in SNR (dB) is defined as

$$\text{Segmental Gain (dB)} = \text{Segmental SNR (Output)}$$
$$- \text{Segmental SNR (Input)}$$
$$= \frac{1}{Mp_1} \sum_{i=1}^{M} \sum_{j=1}^{p_1} \left\{ 10 \, log_{10} \frac{\|\mathbf{s}_i\|_2^2}{\|\mathbf{s}_i - \hat{\mathbf{s}}_i\|_2^2} - 10 \, log_{10} \frac{\|\mathbf{s}_i\|_2^2}{\|\mathbf{n}_i\|_2^2} \right\},$$
$$= \frac{1}{Mp_1} \sum_{i=1}^{M} \sum_{j=1}^{p_1} 10 \, log_{10} \frac{\|\mathbf{n}_i\|_2^2}{\|\mathbf{s}_i - \hat{\mathbf{s}}_i\|_2^2} , \tag{6.15}$$

where $M = 2$ (i.e. representing the stereophonic situations), $\mathbf{s}_i = [s_i(k), s_i(k+1), \ldots, s_i(k + N_f - 1)]^T$, $\hat{\mathbf{s}}_i = [\hat{s}_i(k), \hat{s}_i(k+1), \ldots, \hat{s}_i(k + N_f - 1)]^T$, $\mathbf{n}_i = [n_i(k), n_i(k+1), \ldots, n_i(k+N_f-1)]^T$, $(k = (j-1)N_f, (j-1)N_f+1, \ldots, jN_f - 1, \ j = 1, 2, \ldots, p_1)$ are respectively the clean speech, enhanced speech, and the noise signal vector, and where N_f is the number of the samples in each frame and p_1 is the number of the frames.

The averaged cepstral distance d_{cep} is given by

$$d_{cep} = \frac{1}{M} \sum_{i=1}^{M} \frac{1}{p_{2,i}} \sum_{j=1}^{p_{2,i}} \sum_{k=1}^{2q} (c_{i,k}(j) - c'_{i,k}(j))^2, \tag{6.16}$$

where $c_{i,k}(j)$ and $c'_{i,k}(j)$ are the cepstral coefficients corresponding to the clean and the enhanced signal at left/right channel, respectively. The parameter q is the order of the model, and $p_{2,i}$ $(i = 1, 2)$ is the number of frames where speech is present. The determination of speech presence was achieved

---

[3]Imagine the situation where both the input and output SNRs are high (at 10dB and 22dB for the input and output SNR, respectively). Then, the conventional segmental SNR cannot fully explain how much noise reduction we actually gain, if the input SNR varies greatly (from 5dB to 20dB, say). Hence, we instead consider the segmental gain in SNR as a measurement for noise reduction, since this measurement is also dependent upon the input SNR. (However, in real situations, we normally cannot know both the input and output SNRs.) Nevertheless, in general, the quality of speech enhancement does not always match the subjective evaluation (i.e. the listening tests) and thus, how to evaluate objectively the enhancement quality remains still an open issue (see e.g. Deller et al., 1993).

by manual inspection of the clean speech signals. (Note that normally the numbers of the frames $p_1 \neq p_{2,i}$.)

Figure 6.7 (a) shows a comparison of the segmental gain (given by (6.15)) versus input SNR, for the single stage SSP (with the data-reusing scheme defined in (6.8))[4]. (For the case without the data-reusing scheme, similar results to Fig. 6.7 were obtained. See Hoya et al. (2003b); Hoya et al. (2005) for the detail.)

The results shown in Fig. 6.7 are those averaged over three different speech samples (collected from the corresponding number of speakers, including both the female and male native speakers) of the same sentence used for the simulation examples given so far. In the figure, the performance of the three different noise reduction algorithms, i.e. 1) SSP (using only an SSP), 2) SSP+DASE (i.e. the combination of an SSP and DASE), and 3) NSS algorithm, is compared. For the NSS algorithm, since the performance dramatically varies with the parameter setting (i.e. the NSS algorithms generally have a large degree of freedom in the parameter setting, as described later), three different parameter settings were attempted (indicated as NSS1, NSS2, and NSS3 in Fig. 6.7).

In the figure, at lower SNRs, the performance with NSS is better than the other two, whilst at higher SNRs the SSP+DASE algorithm is the best. However, at lower SNRs, as in Fig. 6.7 (a), the performance in terms of cepstral distance with NSS (for all the three parameter settings) is poorest amongst the three. As in Fig. 6.7 (a), at around SNR > 5(dB), it is clearly seen that the combination of the SSP and DASE yields performance improvement of more than 3(dB) over the case using only the SSP. As the performance improvement of SSP together with the DASE approach observed in Figs. 6.7 (a) and (b) compared to that of only using a single stage SSP, the enhanced signal obtained after the DASE is much closer to the original stereophonic speech signal than that after the SSP.

To see intuitively how the stereophonic image in the enhanced signals by SSP+DASE can be recovered, the scatter plots are shown in Fig. 6.8, where the parameter settings are all the same as those for Fig. 6.5.

In Fig. 6.8(e) (in the figure the labels "ss_1" and "ss_2" correspond to $\hat{s}_1$ and $\hat{s}_2$, respectively, whereas those "nss_1" and "nss_2" correspond respectively to the enhanced signals obtained by the NSS method), it is observed that the pattern of the scatter plot for the enhanced speech after the SSP+DASE somewhat approaches that of the original stereophonic speech as in Fig. 6.8(a), in comparison with that for the speech obtained by applying only the (single stage) SSP shown in Fig. 6.8(d) is considered as rather monaural (which also agreed with the informal listening tests), since the distribution of the data points are more concentrated around the line $s_1 = s_2$ than the case of SSP+DASE.

---

[4]For the computation of segmental gain given in (6.15), the setting $N_f = 256$ was used, whereas $q = 8$ for the cepstral distance in (6.16). Note that the number of frames p_1 and $p_{2,i}$ normally varies with the speech data.

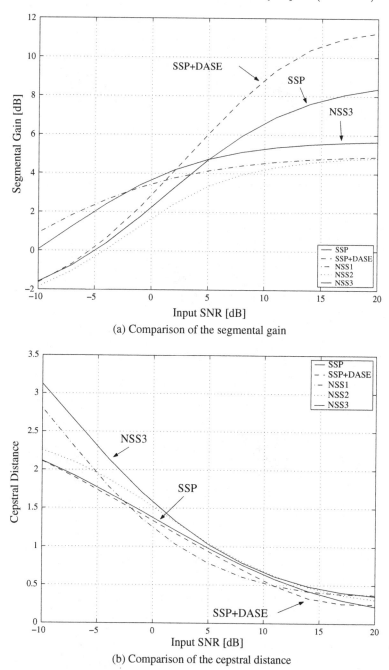

(a) Comparison of the segmental gain

(b) Comparison of the cepstral distance

Fig. 6.7. Performance comparison using the two objective measurements, segmental gain and cepstral distance – averaged over the results using three different speech samples (obtained by applying the data-reusing scheme in (6.8))

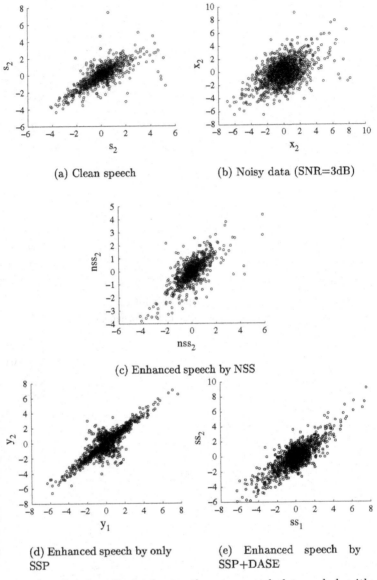

(a) Clean speech

(b) Noisy data (SNR=3dB)

(c) Enhanced speech by NSS

(d) Enhanced speech by only SSP

(e) Enhanced speech by SSP+DASE

Fig. 6.8. The scatter plots obtained using the same speech data and algorithms as in Fig. 6.5

In Fig. 6.8(c), it is also observed that some data points in the original signals are lost (especially at lower-left corner) and that the shape of the cluster is somewhat altered in the enhanced signal by the NSS. This coincides with the empirical fact that the enhanced speech by the NSS can be greatly changed in shape.

6.2.4 Other Studies Related to Stereophonic Noise Reduction

As described earlier, noise reduction has been an active area of research in speech enhancement in the last few decades.

In widely used NSS methods (Martin, 1994; Xie and Van Compernolle, 1996; Gustafsson et al., 1999; Martin, 2001; Gustafsson et al., 2003), both the speech and noise spectra of the noisy speech data are independently estimated by using sample statistics obtained over some number of frames, and then noise reduction is performed by subtracting the spectrum of the noise from that of the observed data. Due to the block processing based approach, however, it is well known that such methods introduce the undesirable musical tonal noise in the enhanced speech, as observed in the simulation examples. As observed in Figs. 6.5 and 6.6, such methods also remove some speech components in the spectra which are fundamental to the intelligibility of the speech in many cases. This is a particular problem at lower SNRs. Moreover, the performance is also quite dependent on the choice of many parameters, such as spectral subtraction floor, over-subtraction factors, or over-subtraction corner frequency parameters. To find the optimal choice of these parameters in practice is therefore very difficult.

Recently, in the study of blind signal processing, one of the most active and potential application areas has been speech separation (Haykin, 2000), and a number of methods for blind separation/deconvolution of speech have been developed (Jutten and Herault, 1991; Nguyen Thi and Jutten, 1995; Torkkola, 1996; Cichocki and Amari, 2002). These methods work quite well, as long as each sensor is located close to each source. However, separation of the speech from noise is still difficult when all the sensors are located close to one dominant source but far from the others, as in cocktail party situations. This sensor configuration is typically employed in practice, for example, as in stereo conferencing systems; two microphones being placed in parallel to each other in front of the speaker at a reasonable distance. Moreover, the existing blind separation/deconvolution methods quite often fail to work where there are more sources than sensors.

In contrast, in the study of biomedical engineering, it has been reported that the utility of the subspace method implemented using the singular value decomposition (SVD) is to successfully enhance nonaveraged data (see e.g. Karjalainen et al., 1999; Kobayashi and Kuriki, 1999). In the technique, the space of the observed data is partitioned into both the signal and noise subspaces. Elimination of the noise is thereby achieved by orthonormal projection of the observed signal onto the signal subspace, with the assumption that the signal and noise subspaces are orthogonal.

In recent studies, a number of SVD based methods have also been developed for speech enhancement (Dendrinos et al., 1991; Ephraim and Trees, 1995; Jensen et al., 1995; Hansen, 1997; Asano et al., 2000; Doclo and Moonen, 2000, 2002). For instance, a Toeplitz (or Hankel) structured data matrix representation is employed within the subspace decomposition operation, and thereby the data matrix is decomposed into signal-plus-noise subspace and a noise subspace rather than signal and noise subspaces (see Ephraim and Trees, 1995; Jensen et al., 1995; Doclo and Moonen, 2002). However, little attention has generally been paid to the extension to multichannel outputs.

As described so far, the combined multi-stage SSP and ASE method works even with the small number of sensors $M = 2$, whilst the conventional approaches such as adaptive beamforming (Howells (1976), see also Haykin (1994) and Hudson (1981)) normally require many number of sensors to function robustly. Moreover, similar to the aforementioned blind separation/deconvolution methods, the traditional adaptive noise cancelling approaches (Widrow et al. (1975), see also Haykin (1994)) have the constraint that one of the sensors must be located close to the noise (or, the reference signal) source. In the scheme described above, such constraint does not exist.

The drawbacks of the combined noise reduction scheme described so far may be the computational complexity required for the SSP part with the order $O(L^3/3)$ (L: length of the analysis matrix), due to the Cholesky's decomposition, which is normally used for the computation of EVD (for more information, see e.g. Golub and Van Loan, 1996), and thus how to efficiently estimate the subspaces still remains an open issue. Nevertheless, this could be relaxed by exploiting on-line subspace estimation approaches (see e.g. Badeau et al., 2004).

6.3 Perception – Defined as the Secondary Output of the AMS

As in Fig. 5.1 (on page 84), the perception module is defined as the secondary output of the AMS and differed from the primary outputs. This is since the perceptual outputs are considered to be *intermediate representation* of the processes occurred within the AMS, whilst the primary outputs are all related to real actions, which can physically affect the surrounding environment and/or vary the conditions of the body.

6.3.1 Perception and Pattern Recognition

In the AMS, the utility of the term *perception* is limited in the sense that the AMS performs various pattern recognition tasks when required from the other modules (thus to give the detailed accounts/justifications in terms of philosophical context is beyond the scope), and, therefore, to internally represent the recognition results for a further data processing within the AMS is meant to be "perception".

As in Fig. 5.1 (on page 84), the pattern recognition results are considered to be obtained by accessing the contents of the various **LTM/LTM-oriented** modules, or the **instinct: innate structure** module via the other modules, where necessary. Then, the pattern recognition results are fed back to the

STM/working memory module, for a further data processing (for the detail, see Sect. 8.3). Therefore, the perception module is closely interrelated to the memory modules described in Chap. 8.

In practice, it is more than desirable that the data processing from the LTM/innate structure modules to **perception** (i.e. the **secondary output**) module is quickly done within the AMS; e.g. as soon as the AMS receives the sensory data (or the feedbacks from other modules) by the **sensation** module and sends them to the STM/working memory, it is possible that the AMS can immediately yield the intermediary perceptual outputs, which in general may be represented in the form of a series of pattern recognition results. However, within the AMS context, how actually the sensory data are treated is quite dependent upon the internal states of the STM/working memory and/or the other associated modules, as will be described in later chapters; the sensory data used for further data processing may be differed from the original by such modules, even if the same sensory data are given to the AMS. Thereby, it is also possible that the *timing* to yield the perceptual outputs may be desynchronised with receiving the sensory data.

In addition, it is also considered that in practice the outputs from the perception module (i.e. the module responsible for yielding the pattern recognition results) are not necessarily visible from the external observers for the AMS to actually function. Nevertheless, by having such visible perceptual outputs, it will be convenient for the external observers (e.g. us) to know the internal representations, e.g. for the purpose of tracing the processes occurring within the AMS and investigating the behaviour, which can be quite helpful, e.g. at the developmental stage of AI.

6.4 Chapter Summary

In this chapter, the two modules of **sensation** and **perception** within the AMS have been described.

As described, the sensation module can be regarded as the input module of the AMS, which interacts with the outside world, receives, and eventually encodes the data into those efficiently processed by the other modules within the AMS. In the description of the sensation module, a practical example of pre-processing mechanism, namely the stereophonic noise reduction, in which the humans binaural data processing is modelled, has been focused through extensive simulation examples and the analyses on the results. The topic of noise reduction is considered to be one of the important and fundamental parts of sensation. Although the noise reduction approach given in this chapter has been considered rather within only the scope of pure signal processing, we will soon return to the issue related to the noise reduction within a more general principle of *learning* in the next chapter.

On the other hand, within the AMS principle, the perception module has been defined as the module which generates a series of pattern recognition results and feeds it back to the STM/working memory module for a further data processing amongst the other modules. Then, in Chap. 8, we will focus upon various memory modules within the AMS, which is closely related to both the sensation and perception modules given in this chapter.

7

Learning in the AMS Context

7.1 Perspective

In this chapter, we dig further into the notion of "learning" within the AMS context. In conventional connectionist models, the term "learning" is almost always referred to as merely establishing the input-output relations via the parametric changes within such models, and the parameter tuning is typically performed by a certain iterative algorithm, given a finite (and mostly static) set of variables (i.e. both the training patterns and target signals). However, this interpretation is rather microscopic and hence still quite distant from the general notion of learning, since it only ends up with such parameter tuning, without giving any clear notions or clues to describe it at a macroscopic level, e.g. to explain the higher-order functions/phenomena occurring within the brain (see e.g. Roy, 2000).

Thus, we firstly begin with the consideration of how the general notion of learning can be interpreted in terms of the interactive processes between the various modules within the AMS and outside the world.

Then, it is described that the learning process is referred to as outcome of the interactive processes between the various modules within the AMS and can be eventually ascribed to both the parametric and structural changes within the associated modules. Amongst all such modules, the memory modules play the central role, the modules of which will be described at full length in the next chapter.

7.2 The Principle of Learning

In real life, it is intuitively/naturally considered that not only the sensational and perceptual mechanisms but also other functionalities have been evolved in structure, through generations by generations, in order to adapt themselves and survive in the surrounding environment and varying situations for the continuous existence of the species. Thus, this always involves the

Tetsuya Hoya: *Artificial Mind System – Kernel Memory Approach*, Studies in Computational Intelligence (SCI) **1**, 117–133 (2005)
www.springerlink.com

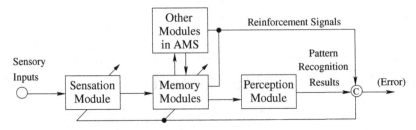

Fig. 7.1. A macroscopic representation of the general evolutionary process in terms of sensation, memory, perception, and other modules within the AMS

interaction between the individuals and the outside world. (In this respect, the concept of learning agrees with the general "feedback" principle (see Wiener, 1948; Simon, 1996).) For instance, it is generally acknowledged that, as a consequence of the adaptation, we human-beings are equipped with five sensors to acquire and process auditory, gustatory, olfactory, tactile, and visual information, so as to interact with the world.

In the previous chapter, it has been described that the sensation module in AMS can be regarded as a combination of multiple sensors and a cascade of the pre-processing units. It is then considered that some of the pre-processing units within the sensation module, as well as the memory modules, must also be evolved, according to the incessantly varying conditions. In this principle, a macroscopic representation of this general evolutionary process for both the sensation and perception modules can be illustrated in Fig. 7.1.

As in Fig. 7.1, after that the AMS receives the sensory inputs from the sensation module, the memory modules process the (encoded) data, which are obtained from a certain number of pre-processing stages within the sensation module, and the perception module eventually yields the pattern recognition results as the secondary outputs of the AMS. Along with this regular data processing, the AMS also performs its self-adaptation to the current situations; in Fig. 7.1, the "C" in the circle denotes a comparator between the perceptual outputs (i.e. pattern recognition results) and reinforcement signals (or target responses) and yields the "error" sequence. This error sequence is then used to re-structure both the sensation and memory modules. (In addition, albeit not explicitly shown in Fig. 7.1, it is implied that, besides sensation and perception, other modules can be re-structured by the interactive processes with the memory modules. Later in the present chapter, this will be discussed further.)

Therefore, it is said that the macroscopic representation in Fig. 7.1 describes the general concept of "reinforcement learning" (Turing, 1950; Minsky, 1954; Samuel, 1959; Mendel and McLaren, 1970). Within the AMS context, the notion of *learning* is accordingly defined, in such a way that the whole system is evolved according to the reinforcement (or, *rewards*). As depicted in Fig. 7.1, the signals for the reinforcement are then regarded as the outcomes of the interactive processes between the memory and/or other modules within the AMS and eventually used to self-evolve the entire system.

In Fig. 7.1, the directional flow from the memory modules to the comparator indicates some possibility that the reinforcement signals for evolving some parts of the memory modules can be given from the others; i.e. imagine a situation that a certain SOKM is responsible for a particular domain of the auditory sensory (or the encoded) data, the reinforcement signals for the SOKM can be eventually given from, say, the other SOKM(s) responsible for, e.g. the corresponding visual counterpart.

7.3 A Descriptive Example of Learning

Now, in order to see more in detail what processes in terms of learning can be involved within the AMS, let us examine a simple example of general learning, imitating a situation that a child is about to learn how to pitch a ball to the targeted point (e.g. a catcher sitting moderately far in front); intuitively, within the AMS context, the following five major steps are then considered to be involved:

Step 1) Perception of the targeted position.

Step 2) Motion planning (or thinking) – for pitching a ball to the targeted position.

Step 3) Performing real (motoric) actions – to actually pitch a ball.

Step 4) Perception of the success/failure.
- If the AMS *recognises* it as a success, go to Step 5).
- Otherwise, go back to Step 2).

Step 5) Updating the contents of (long-term) memory (i.e. "learning by heart").

As depicted in Fig. 7.1, Step 1) in the above involves the following three minor steps: 1) the data processing within the **sensation** module (i.e. performing a series of pre-processing actions to encode the incoming raw data, e.g. encoding the visual sensory data of the targeted position for a further processing within the AMS); 2) the interactive processes with using the encoded data obtained in 1) between the memory modules and other associated modules within the AMS; and eventually 3) the data processing to yield subsequently the perceptual outputs obtained from the **perception** (or **secondary output**) module by accessing the (long-term) memory modules (where appropriate). As described in the previous chapter, the perceptual outputs are generally given in the form of pattern classification results within the AMS context.

Then, in Step 2), motion planning is (mainly) performed via the **thinking** module, in order to pitch a ball so that it reaches the exact position of the target established in Step 1). This planning is mainly carried out on the basis of the three factors: 1) the perceptual outputs of the targeted position obtained in Step 1); 2) the perception (or "recall") of the physical limitations given by the **innate structure**; and 3) the interactive data processing with the other modules attached/functioning in parallel, i.e. the **STM/working memory**, **attention**, **emotion**, or **intuition** module.

Once the planning is completed, in Step 3), the AMS activates the kernel units within the specific areas of the **implicit LTM** (e.g. such as procedural memory, to be described in Sect. 8.4.2), the kernel units of which are directly connected to the **primary output** module, by following (one by one) the planning procedure so established in Step 2) (which can be temporarily represented e.g. as a form of the kernel network within the STM/working memory module), in order to actually perform a series of the motoric actions and pitch a ball towards the target.

In Step 4), it is considered that the perception of success/failure falls in either of the two cases: perception of the success or failure i) *during* performing the motoric actions in Step 3); and ii) *after* that the ball is released from the body. For the former, it is implied that the actual motoric actions being performed are also (somewhat) monitored via the feedback inputs, i.e. the connection between the STM/working memory and primary output modules, as in Fig. 5.1. Then, by comparing, one by one, these feedbacks and a sequence of the planned (or, imagined) tasks during the performance, the AMS can perform the perception of the success/failure. In contrast, the latter implies the perception, if the ball so released can reach/has reached the target, via the sensory data (i.e. the visual/auditory sensory data) so processed. Thus, the manner of perception in both the cases i) and ii) can be a similar one to that performed in Step 1). Nevertheless, it is said that, for both cases, the perception of success/failure therefore strongly depends upon the outcomes of the interactive processes within the STM/working memory and other associated modules.

Lastly, Step 5) involves restructuring/updating the contents of the memory modules to complete the learning process; the resultant perceptual processes performed in Steps 1, 2, and 4) not only yield the sequential activations from a particular set of kernel units but also (temporarily) leave the "trace" of such activations within the STM/working memory module. In other words, some "outstanding" events to the AMS (e.g. the result of the failure in the first trial) occurred during the learning process in Steps 1–4) will remain in the memory for a certain period of time, under the assumption that not all but some events are left within the memory. This is considered, due to e.g. the memory capacity of the STM/working memory or, eventually, even the innate structure (as well as emotion, since it is considered to function in parallel to the innate structure: instinct module) module to influence the "importance" of the events. In the AMS context, such memory trace can then be represented by both the kernel units and their link weights remaining within the STM/working memory and, later, may be eventually transferred to the LTM modules (or, in other words, the events are "imprinted" within the LTM). As will be explained in detail in Sect. 8.3, this involves the interactive data processes between the STM/working memory and **explicit LTM** module.

In summary, the major Steps 1–5) in the above show an example of general learning within the AMS context, whilst Step 5), which results in the formation/restructuring of the LTM modules, can be referred to as (in a narrower sense of) "learning" in terms of the memory context (for a thorough psychological justification, cf. Anderson, 2000).

7.4 Supervised and Unsupervised Learning in Conventional ANNs

In conventional ANNs, the manner of "learning" can be classified into two categories, i.e., supervised and unsupervised learning. In the supervised learning scheme, the target response (or, the teacher signal) is normally required, and in practice such response is artificially given/pre-determined and only used to establish the input-output mappings (e.g. the mappings between the training and target responses data given) by such networks. Thus, the learning is, in a strict sense, not autonomous at all, since the target values are normally pre-determined by humans, and hence the utility of the term "learning" is restricted. In contrast, within the unsupervised scheme, though this sort of target responses is not necessary, the manner of the mapping construction using ANNs is usually quite dependent upon the statically-given training data set (i.e. in the statistical context, and also, the order of presenting the training patterns often affects the performance in e.g. constructive approaches). Moreover, within the ordinary learning schemes in the ANN context, the distinction between the data used and those not desirable for the training, such as the so-called "noisy" data or "outliers", is normally made only in a strict statistical sense.

However, it is considered that such distinction would be only effective to e.g. achieve a (statistical) function approximation or construct a static pattern classifier, but ineffective to develop a more dynamic scheme; at some time, the data can be regarded as simple noise, but some other time(s), even they are treated oppositely, i.e. as the information of interest for an intelligent mechanism. Then, the decision must be quite dependent upon e.g. the internal states/innateness of the whole system (e.g. due to the **emotion** module, to be described in Chap. 10). For instance, imagine a situation that we would like to realise the mechanism, as in the descriptive example in Sect. 5.2.2, that a specific voiced input, i.e. the (encoded) auditory data of a friend's whispering, can be regarded as simple noise, whilst we are attentive to the orchestral sound data (see also (in p.7) Minsky, 1979).

7.5 Target Responses Given as the Result from Reinforcement

As described above, in conventional ANNs, the target responses within the supervised learning scheme are in practice given by humans.

In contrast, these externally given or "straightforwardly provided" target values are generally not considered within the AMS context; even if the (so-called) target responses are presented to the AMS (e.g. by humans), such responses are firstly received by the sensation module, then the perceptual outputs are fed back to the STM/working memory module via the LTM modules, processed by the associated modules, and eventually used for performing the evolutionary process (or, in other words, the "self-evolutionary" process by the AMS) as illustrated in Fig. 7.1. Thus, it is not always true, as we humans, that these "target" responses given externally are processed and used in the original forms within the AMS; the responses may be "coordinated" (or modified) by some associated modules (and in due course even the outcome may be completely different from those originally presented to the AMS), e.g. due to the strong influence by the innate structure (or the "instinct" module) of the AMS (to be described in Sect. 8.4.6) which defines the "life values" (or the preset values e.g. inherent to the physical limitation of the system). Then, it is said that how to deal with the target responses given externally depends upon the resultant internal processes occurring within the AMS, unlike conventional ANN schemes. (In this principle, the term "reinforcement" is more appropriate within the AMS than the "target responses".)

In conventional GRNNs/PNNs, to "learn" adds/removes the RBFs from the network and then simply assigns the target values to the corresponding weights between the neurons in the hidden and output layers (as described in Sect. 2.3.1). In this sense, it is also seen that the latter sense of learning, i.e. the assignment of the target values is equivalent to establishing the connections between the two distinct layers. Then, generalising this leads to learning in the AMS context[1]; in kernel memory, since lateral connections are allowed, such distinct layers do not exist, and, ultimately speaking, "learning" is attributed to establishing the connections between the kernel units (as well as adding/removing kernel units, where appropriate) and set the link weight values in between, *after the interactive processes between various modules.*

In general pattern classification tasks, the learning is hence meant to be the establishment of the connections between the kernel units representing the objects and those representing the target values. For the kernel units representing such target values, either ordinary or symbolic kernel units (i.e. the latter with the kernel function given as (3.11)) may be exploited.

Then, it is intuitively considered that such target value (or teacher/the reinforcement signal) can be given from the kernel unit(s) which has successfully

---

[1]Note that, as described in Chap. 3, there are no strict sense of layers defined within the kernel memory concept, unlike in ordinary GRNNs/PNNs, and that the connections between the kernel units and those representing the output nodes can

remained, e.g. within a certain SOKM for a relatively long period of time (i.e. LTM), yields consistently the proper activations by the appropriate stimuli, and thus may have more connections (or associations via the link weights with other kernel units) than the newly added kernel unit(s). (Thus, this notion somewhat resembles a scenario that a "teacher" provides students some directions, in order to give them the opportunities to expand their knowledge, or "associate"/"link" their already acquired knowledge with other matters).

In other words, during the learning process of memory modules, though initially the SOKMs responsible for the respective domains/modalities are formed rather in an unsupervised manner, they in contrast can be consolidated rather in a supervised manner in the later process, due to e.g. the reinforcement signal(s) given from a cluster of the kernel units (and/or the (symbolic) kernels representing the class labels) formerly formed in other modalities.

7.6 An Example of a Combined Self-Evolutionary Feature Extraction and Pattern Recognition Using Self-Organising Kernel Memory

In conventional approaches, the topics of feature extraction and pattern recognition have normally been dealt with separately from each other. In feature extraction, the raw data received by sensors are generally encoded into some other forms (or "patterns") to be relatively conveniently handled by the post-processors. In contrast, in pattern recognition, classification of the patterns obtained after the feature extraction process is actually performed by the (so-called) post-processors; i.e. the "pattern classifiers".

Here, as an example of the reinforcement learning, we consider a model of combined feature extraction and pattern classification using the self-organising kernel memory (in Chaps. 3 and 4) that can self-evolve according to the time-varying situations. Without loss of generality, we here limit ourselves to consider that the reinforcement signals are already given.

Figure 7.2 shows the block diagram of a combined self-evolutionary feature extraction and pattern recognition system. As in Fig. 7.2, the system consists of the five units; i.e. 1) the sampling unit[2]; 2) the subband coding unit; 3) the unit to form the input data for SOKMs; 4) that consisting of the SOKMs; and 5) that generating the reinforcement signals (or target responses) to restructure (or perform the reinforcement learning of) the units 1)-4), with respect to the error between the pattern recognition results and

---

be varied without affecting the contents of the memory (stored as a form of the template vectors) within the kernel functions.

[2]In Fig. 7.2, without loss of generality, a digital system is assumed, with the principle that normally in practice the data can be more conveniently/efficiently handled by the current digital systems than analog ones. However, it can also be possible to generalise this diagram for analog systems.

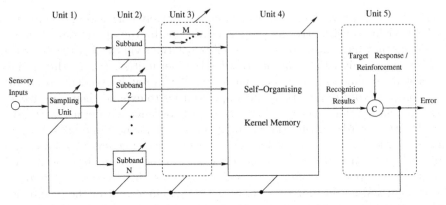

Fig. 7.2. Block diagram of a combined self-evolutionary feature extraction and pattern recognition system – consisting of the five units: 1) the sampling unit; 2) the subband coding unit; 3) the unit to form the input data for SOKMs; 4) that consisting of the SOKMs; and 5) that generating the reinforcement signals (or target responses) to restructure the units 1)-4), with respect to the error between the pattern recognition results and the target values obtained from the comparator (denoted in "C" in the circle)

the target values obtained from the comparator (denoted by "C" in the circle, corresponding to that in Fig. 7.1 (on page 118)). Then, the first four units 1)-4) will be evolved in their structure by the reinforcement learning.

In the AMS context, it can be seen that the units 1)-3) belong to the **sensation** module, whereas the units 4) and 5) involve the memory modules (as well as the **innate: instinct** module) and both the **primary** and **secondary: perceptual output** modules.

Now, we consider each of the five units in more detail in the following three subsections.

7.6.1 The Feature Extraction Part: Units 1)-3)

In Fig. 7.2, it is considered that the subband unit 2) can be represented by a bank of bandpass filters (e.g. realised by the approach using quadrature mirror filters (QMFs)) (see e.g. Crochiere and Rabiner, 1983; Deller et al., 1993) or the time-frequency analysis with the utility of wavelets (see e.g. Mallat, 1999)[3], etc. For the feature extraction part in Fig. 7.2, it is assumed that the sampling rate in Unit 1) and both the number of subbands and parameters for the respective filter banks (i.e. the filter coefficients to determine e.g. the pass, transition, or stop-band) in Unit 2) can be varied.

Note that, as aforementioned in Sect. 6.2.1, a subband structure can in general be regarded as the pre-processing mechanism, which is universal to

---

[3]In the case of exploiting a time-frequency analysis, both Units 2) and 3) can be represented at a time.

the sensory modality in humans such as auditory, visual, or olfactory, and thus that the feature extraction part in Fig. 7.2 could also be universal to describe the models for the other two modalities: gustatory and somatosensory. In a more engineering context, we even may (and in practice we do) exploit further the subband structure to the pre-processing of both the biomedical data, e.g. electrocardiography (ECG), electromyography (EMG), or the brain wave related representations, such as electroencephalography (EEG), functional magnetic resonance imaging (fMRI), magnetoencephalography (MEG), positron emission tomography (PET), or single-photon emission computed tomography (SPECT), and non-biomedical signals, e.g. communication, radar, seismic, or sonar sensory signals.

In Unit 3), the subband data obtained from Unit 2) will be collected (in time-wise or, in other words, in the form of "frames") data and eventually sent to Unit 4) as the feature data for the post pattern classifiers. Thus, during the reinforcement learning process performed by Unit 5), the parameter which determines the number of frames (and, if appropriate, those in some other functions to form the feature data) can also be varied.

7.6.2 The Pattern Recognition and Reinforcement Parts: Units 4) and 5)

In contrast to Units 1)-3), Unit 4) actually performs the pattern recognition, consisting of several (sub-)SOKMs, the structure of which will be varied, during the reinforcement learning process, and eventually sends the pattern recognition results obtained from such sub-SOKMs to Unit 5).

Now, let the subband data obtained from Unit 2) at time index n (i.e. for a single frame) be

$$\mathbf{x}(n) = [x_1(n), x_2(n), \ldots, x_N(n)]^T \ , \qquad (7.1)$$

then, after the data arrangement in Unit 3), the input data to the sub-SOKMs $\mathbf{Y}(n)$ can be written (in a matrix form):

$$\mathbf{Y}(n) = f([\mathbf{x}(n), \mathbf{x}(n-1), \ldots, \mathbf{x}(n-M+1)]) \ , \qquad (7.2)$$

where the function $f(\cdot)$ can be given 1) to smooth the envelope further at each frame (i.e. in a row-wise operation), 2) to quantise the data further in time-wise (i.e. in a column-wise operation), and/or 3) to normalise the values of the data (i.e. in both the row and column-wise operation), e.g., as aforementioned, for keeping a "well-balanced" set of data points for the pattern space. Thus, the size of the data matrix $Y(n)$ can also be varied to $(N' \times M')$, where $N' \leq N$ and $M' \leq M$.

In Unit 4), it is considered that, although the sub-SOKMs in Fig. 7.2 are considered to be responsible for a single modality (e.g. a certain auditory sensory data), they can be configured differently from each other; i.e. each sub-SOKM consisting of the kernel units (and the connections via the link weights) with having their template vectors (defined in a different dimensionality) to represent other modality.

Then, during the learning process, the respective sub-SOKMs can be reconfigured within the so-called *competitive learning* principle (for the general notion, see von der Malsburg, 1973)[4], to be described later.

7.6.3 The Unit for Performing the Reinforcement Learning: Unit 5)

As aforementioned, Unit 5) sends the reinforcement signals to reconfigure the units 1)-4). In this example, for simplicity, it is assumed that the reinforcement signals are given, i.e. based upon the statistics of the errors between the pattern recognition results and externally provided (or pre-determined) target responses, as in ordinary ANN approaches. (In such a case, the comparator denoted by "C" in the circle in Fig. 7.2 can be replaced with a simple operator that yields the error.) However, within a more general context of reinforcement learning as described in Sect. 7.5, the target responses (or reinforcement signals) can be given as the outcome from the interactive processes between the modules within the AMS.

7.6.4 Competitive Learning of the Sub-Systems

Without loss of generality[5], as shown in Fig. 7.3, consider that the combined self-evolutionary feature extraction and pattern recognition system, which is responsible for a particular domain of sensory data (i.e. for a single category/modality), consists of the two (partially distinct) sub-systems A and B.

Then, suppose that the respective feature extraction (i.e. Units 1)-3)) and pattern classification parts (i.e. Unit 4) are configured with two distinct parameter sets A and B; i.e. both feature extraction A and sub-SOKM A have been configured with parameter set A during a certain period of time p_1, whereas both feature extraction B and sub-SOKM B have been formed with parameter set B during the period p_2, and that both the sub-systems are working in parallel.

Based upon the error generated from the comparator C_1 (attached to both the sub-SOKMs A and B), the comparator C_2 within Unit 5) yields the signals to perform the competitive learning for sub-system A and B; i.e. firstly, after the formation of the two sub-systems in the initial periods p_1 and p_2,

---

[4]Note that, unlike in ordinary ANNs context (e.g. Rumelhart and Zisper, 1985), here the terminology "competitive learning" is used in the sense that the competitive learning can be performed at not only neuronal (i.e. kernel unit) but also system levels within AMS.

[5]The generalisation for the cases where there are more than two sub-systems is straightforward.

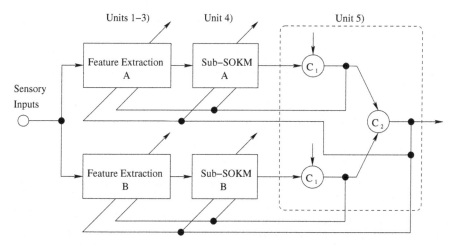

Fig. 7.3. An example of competitive learning within the self-evolutionary feature extraction and pattern recognition system – two (partially distinct) sub-systems A and B reside in the system

the statistics of the error between the reinforcement signals (target responses) given and pattern classification results for both the sub-systems A and B will be taken during a certain period p_3. Then, on the basis of the statistics taken during the period p_3, if the error rates obtained from sub-system A are higher than those from sub-system B, for instance, only sub-system A can be intensively evolved (i.e. some of the parameters within the units 1)-4) of sub-system A can be varied greatly), whilst sub-system B is (almost) fixed, with only allowing some small changes in the parameter settings which do not give a significant impact upon the overall performance[6], during the subsequent period of time p_4. Similarly, this process is repeated endlessly, or e.g. until reasonable pattern classification rates are obtained by either of the two sub-systems. Figure 7.4 illustrates an example of the time-course representation of this repetitive process.

Moreover, it is also considered that, if either of the two does not function well (e.g. the classification rates have been below or the number of kernel units activated has not reached a certain threshold for several periods of time), the complete sub-system(s) can be eventually removed from the system (i.e. representing *"extinction"* of the sub-system).

---

[6]For instance, suppose that the sub-SOKM in Unit 4) has a sufficient number of kernel units to span a pattern space for a particular class, a small change in the number of kernel units would not cause a serious degradation in terms of the generalisation capability (see Chaps 2 and 4, for more practical justifications).

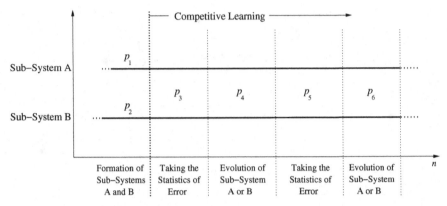

Fig. 7.4. An example of the time-course representation of the competitive learning process – here, it is assumed that the system has two sub-systems A and B, configured respectively with distinct parameter sets A and B. Then, after the formation of both the sub-systems (during the period p_1 for sub-system A and p_2 for sub-system B), the competitive learning starts; during the period p_3 (p_5), the statistics of the error between the reinforcement signals (or target responses) and pattern classification results (due to the comparators in Unit 5) are taken for both the sub-systems A and B, then, according to the error rates, either of the two sub-systems will be intensively evolved during the next period p_4 (p_6). This is repeatedly performed during the competitive learning

7.6.5 Initialisation of the Parameters for Human Auditory Pattern Recognition System

In Units 1)-3), it is considered that the following five parameters can be varied:

i) Sampling frequency: f_s (in Unit 1)
ii) Number of subbands: N (in Unit 2)
iii) Parameters for designing the respective filter banks (in Unit 2)
iv) Number of frames: M (in Unit 3)
v) Function: $f(\cdot)$ (in Unit 3) and (if appropriate) the internal parameter(s) for $f(\cdot)$

whereas the parameters for the sub-SOKMs in Unit 4), as given in Table 4.2, can also be varied, during the self-evolutionary (or the reinforcement learning) process for the system.

Then, if we consider an application of the self-evolutionary model described earlier to develop a self-evolutionary human auditory pattern recognition system, the initialisation of the parameters can be done, by following the neurophysiological/psychological justifications of human auditory perception (Rabiner and Juang, 1993; Warren, 1999), and thereby the degrees of freedom can, to a great extent, be reduced in the parameter settings and/or the competitive learning process can be accelerated.

For instance, by simulating both the lower and upper limit of the frequency range (normally) perceived by humans, i.e. the range from 20 to 20,000Hz, the first three parameters, i.e. i) f_s (the sampling frequency in Unit 1)), ii) N (the number of subbands), and iii) the parameters for designing the respective filter banks in Unit 2), can be determined *a priori*.

For iii), a uniform filter bank (Rabiner and Juang, 1993) can be exploited, for instance. Alternatively, the utility of nonuniform filter banks with mel or bark scale can immediately specify the parameters ii) and iii) in Unit 2), in which the spacings of filters are given on the basis of perceptual studies, and can be generally effective in speech processing, i.e. to improve the classification rates in speech recognition tasks.

On the other hand, the fourth parameter, i.e. the number of frames, M may be set, with respect to e.g. the retention of memory in the STM, which has been well-studied in psychology (Anderson, 2000).

In general speech recognition tasks, the fifth $f(\cdot)$ can be appropriately given as a combined smoothing envelope and normalisation function. For representing the former function, a further quantisation of data is performed (i.e. resulting in smoothing the envelope in each subband e.g. by applying a lowpass filter operation), whilst the latter is normally used in conventional ANN schemes, in order to maintain the well-spanned data points of a feature vector in the pattern space (by the ANNs).

In the self-evolutionary pattern recognition system, such settings as in the above can be effectively used to initialise all the five parameters i)-v), and, where appropriate, some of those in i)-v) can be reset, according to the varying situations. This can thus lead to a significant reduction in computation to reach a "steady state" of the system, as well as decrease in the degrees of freedom within the initial parameter settings, for performing the self-evolutionary process.

In a similar fashion to the above, the initialisation of the parameters i)-v) can be achieved for other modalities.

7.6.6 Consideration of the Manner in Varying the Parameters i)-v)

As described in the above, the degrees of freedom in the combined self-evolutionary feature extraction and pattern recognition system can be large. Here, we consider how the system can be efficiently evolved during the learning process, from the aspect of varying the parameters.

It is intuitively considered that the feature extraction mechanism, i.e. that corresponding to the subband coding in Unit 2) or the formation of the input data to the sub-SOKMs by Unit 3) as in Fig. 7.2, can be (almost) seen as a static mechanism (or, if any, may be evolved in a extremely "slow" pace, i.e. evolved through generations by generations), within both the principles in human auditory perception (see e.g. Warren, 1999) and the retention of memory in STM (Anderson, 2000). In contrast, the pattern classification mechanism can be rather regarded as more "plastic" and thus evolve faster than the feature extraction counterpart.

From these postulates, it may therefore be said that in practice varying the parameters i)-iv) can give more impact upon the evolutionary process (as well as the overall performance) than those by the other parameters in relation to the pattern classifiers (i.e. the sub-SOKMs).

Within this principle, the parameters inherent to the self-evolutionary system could be varied, according to the following periods of time:

In period q_1): Varying the parameters with respect to the sub-SOKMs (Unit 4)

In period q_2): Varying (if appropriate) the internal parameters for $f(\cdot)$ (Unit 3)

In period q_3): Varying the number of frames M (Unit 3)

In period q_4): Varying the number of subbands N and the designing parameters for the filter banks (Unit 2)

In period q_5): Varying the sampling frequency f_s (Unit 1)

where $q_1 < q_2 < \ldots < q_5$.

Then, where appropriate, the parameters may be updated by e.g. the following simple strategy:

$$v = \begin{cases} v_{min} & \text{; if } v < v_{min} , \\ v_{max} & \text{; else if } v > v_{max} , \\ v + \delta_v & \text{; otherwise} , \end{cases} \tag{7.3}$$

where v corresponds to one of the parameters related to the self-evolutionary system, v_{min} and v_{max} denote the lower and upper bound, respectively, which may be determined *a priori*, by taking into account e.g. the physical limitations inherent in each constituent of the system, and δ_v is either a negative or positive constant.

7.6.7 Kernel Representation of Units 2)-4)

As aforementioned, in Unit 2) (and Unit 3), a subband coding can be performed by "transforming" the raw data into another domain (e.g. time-frequency representation) for conveniently dealing with the data by the post processors/modules within the AMS. As postulated in the neurophysiological study (Warren, 1999), processing the sound data in human auditory system begins with the subband coding similar to the Fourier analysis for which both the basilar membrane and inner/outer cells within the cochlea of both the ears are responsible.

We here consider that the subband coding processing can also be represented within the kernel memory principle:

The first half of the discrete Fourier transform (DFT) of a signal sequence $\mathbf{x} = [x_1, x_2, \ldots, x_L]$ (i.e. with finite length $L = 2N$) X_i ($i = 1, 2, \ldots, N$) is given by (see Oppenheim and Schafer, 1975)

$$X_i = \sum_{k=0}^{L-1} x_k W_L^{ik}$$

$$W_L = \exp\left(-j\frac{2\pi}{L}\right) \tag{7.4}$$

where W_L is a Fourier basis.

Now, using the inner product representation of the kernel function in (3.4), the Fourier transform in (7.4) can be redefined as a cluster of N kernel units with the respective kernel functions K_i^ϕ $(i = 1, 2, \ldots, N)$[7]:

$$K_i^\phi(\mathbf{x}) = \mathbf{x} \cdot \mathbf{t}_i \tag{7.5}$$

where each template vector $\mathbf{t}_i$ is given as a collection of the Fourier bases:

$$\mathbf{t}_i = [t_1^i, t_2^i, \ldots, t_L^i]^T \ ,$$
$$t_k^i = W_L^{i(k-1)} \ (k = 1, 2, \ldots, L) \ . \tag{7.6}$$

Note that, with the representation in (7.5), each kernel unit K_i^ϕ can be seen as a distance metric for the i-th frequency bin, by comparing the input data with its template vector given by (7.6).

Then, Fig. 7.5[8] shows another representation of Units 2)-4) within only the kernel memory principle. As in the figure, alternative to the subband representation in (7.2) for Unit 3), the matrix

$$\mathbf{Y}(n) = f([\mathbf{y}(n), \mathbf{y}(n-1), \ldots, \mathbf{y}(n-M+1)]) \ (\in \Re^{N' \times M'})$$
$$\mathbf{y}(n) = [K_1^\phi(\mathbf{x}(n)), K_2^\phi(\mathbf{x}(n)), \ldots, K_N^\phi(\mathbf{x}(n))]^T \tag{7.7}$$

can be given as the input to the kernel units within sub-SOKMs A-Z, where the function $f(\cdot)$ is the same one used in (7.2).

Note that the representation for other transform(s), such as discrete sine/cosine or wavelet transform, can be straightforwardly made within the kernel memory principle.

7.7 Chapter Summary

This chapter has focused upon the concept of learning and its redefinition within the AMS context. As described in this chapter, the term "learning"

---

[7]Here, it is assumed that the kernel function can deal with complex values, which can be straightforwardly derived from the expression in (3.2). Nevertheless, since the activation of such kernel unit can always be represented by a real value(s), this does not affect other kernel units connected via the link weights at all.

[8]In Fig. 7.5, each sub-SOKM in Unit 4) is labeled with the superscripts from A to Z and arranged in an alphabetic order for convenience. However, this manner of notation does not imply that the maximum number of sub-SOKMs is limited to 26 (i.e. the total number of the alphabets A-Z).

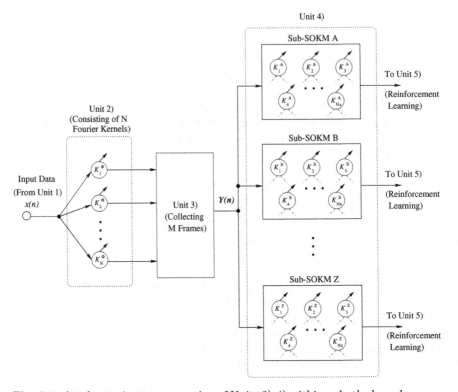

Fig. 7.5. An alternative representation of Units 2)-4) within only the kernel memory principle; Units 2)-4) consist of both N Fourier kernel units (in Units 2) and 3)) and the sub-SOKMs (A-Z) (in Unit 4). Eventually, the output from each sub-SOKM is fed into Unit 5) for the reinforcement learning process

appeared in most conventional connectionist models merely specifies the parameter tuning to achieve the input-output mapping, given both the training patterns and target responses, and hence, the utility of the term is quite limited. Moreover, in such models, the target responses are usually predetermined by humans.

In contrast, within the AMS context, a more general notion of learning and the target responses has been redefined, by examining a simple example of learning. For performing the learning process by AMS, it has been described that various modules within the AMS, i.e. attention, emotion, innate structure, the memory modules, i.e. the STM/working memory and explicit/implicit LTM, perception, primary output, sensation, and thinking module, are involved.

Then, an example of how to construct a self-evolutionary feature extraction and pattern recognition model in terms of the AMS has been given. In practice, such a combined approach can be applied to the so-called "data-mining", in which some useful components can be automatically extracted from the raw data (though, in such a situation, the performance is considered to be heavily dependent upon the sensory part of the mechanism). On the other hand, it is considered that the appropriate initialisation of the parameters, i.e. for the sensation mechanism, can greatly facilitate the evolution processing. For this, the *a priori* knowledge of the human sensory system and how to implement it during the design stage of the self-evolutionary model can be of fundamental significance. In addition, it has been described that some parts within the self-evolutionary model can be alternatively represented by the kernel memory.

In the following chapter, the memory modules within the AMS, which are closely tied to the notion of learning, will be described in more detail.

8

Memory Modules and the Innate Structure

8.1 Perspective

As the philosopher Miguel de Umamuno (1864-1936) once said,

> "We live in memory and memory, and our spiritual life is at bottom simply the effort of our memory to persist, to transform itself into hope ... into our future."

> from "Tragic Sense of Life" (Unamuno, 1978),

the "memory" is an indispensable item for the description of the mind. In psychological study (Squire, 1987), the notion of "learning" is defined as the process of acquiring new information, whereas "memory" is referred to as the persistence of learning in a state that can be revealed at a later time (see also Gazzaniga et al., 2002) and the outcome of learning. Thus, both the principles of learning, as described in the previous chapter, and memory within the AMS context are closely tied to each other.

In this chapter, we focus upon various memory and memory-oriented modules in detail, namely the 1) **STM/working memory**, both 2) **explicit (declarative)** and 3) **implicit (nondeclarative) LTM** modules, 4) **semantic networks/lexicon**, and 5) the **innate structure** (i.e. pre-defined architecture) within the AMS, as well as their associated interactive data processing with the other modules. It is then described that most of the memory-oriented modules within the AMS can be realised within a single framework of the kernel memory given in the previous Chaps. 3 and 4.

8.2 Dichotomy Between Short-Term (STM) and Long-Term Memory (LTM) Modules

As in Fig. 5.1 (on page 84), the memory modules within the AMS are roughly divided into two types; the short-term/working and long-term memory modules, depending upon the i) retention, ii) capacity to store the information (in

Tetsuya Hoya: *Artificial Mind System – Kernel Memory Approach*, Studies in Computational Intelligence (SCI) **1**, 135–168 (2005)
www.springerlink.com

the form of encoded data) within the kernel units, and iii) the functionality, the division of which directly follows the cognitive scientific/psychological memory dichotomy (James, 1890). In the AMS context, the STM/working memory is considered to function normally with consciousness (but at some other times subconsciously), whereas the LTM modules work without consciousness. As described previously (in Sect. 5.2.1), the STM/working memory can be normally regarded as the module functioning consciously in that, where necessary, any of the data processing within the STM/working memory can be mostly directly accessible/monitored from other (consciously) functioning modules.

This notion of memory dichotomy between the STM/working memory and LTM is already represented in terms of the memory system in today's Von-Neumann type computers; the main memory within the central processing unit (CPU) resembles the STM/working memory in that a necessary chunk of data stored in the auxiliary memory devices, which generally has much more capacity than the main memory and can thus be regarded as the LTM, are loaded at a time and (temporarily) stay there, for a while, until a certain data processing is completed.

Turning back to the AMS, in practice, the actual (or geometrical) partitioning of the entire memory space, which can be composed by multiple kernel units, into the corresponding STM/working memory and LTM parts, is, however, not always necessary, since it may be sufficient to simply *mark* and hold temporarily the absolute locations/addresses of the kernel units within the memory space, the kernel units of which are activated by the data processing within the STM/working memory, e.g. due to the incoming sensory data arrived from the sensation module. From the structural point of view, the kernel units with a relatively shorter duration of existence can be regarded as those within the STM/working memory module, whereas the kernel units with a longer (or nearly perpetual) duration can be considered as those within the LTM modules. Then, the STM/working memory module also contains e.g. a list relevant to the information about the absolute locations (i.e. the absolute addresses) of the activated kernel units within the entire memory space.

At any rate, for the purpose of simulating the functionality of STM/working memory, it is considered that the issue of which representation is confined to the implementation and thus is not considered to be crucial, within the AMS context.

8.3 Short-Term/Working Memory Module

The STM/working memory module plays the central part for performing the interactive processes between other associated modules within the AMS. In cognitive scientific/psychological studies, it is generally acknowledged that the STM (or working memory) is the "seat" for describing consciousness. (Further discussion of consciousness is left until Chap. 11).

In AMS, since both the functionalities of STM and working memory are rather considered to be complementary to each other, both the notions of STM and working memory can be treated within a single module; the term STM implies relatively short duration of retaining the information, in contrast to the LTM modules; whereas, under the name "working memory", such information can be dealt, or even coordinated/deviated from the original, within the "working memory", due to the interactive processes with the associated modules. Hence, the name "STM/working memory".

Moreover, with respect to the short-term retention of information in memory, it is considered in some studies in cognitive science/psychology (cf. Atkinson and Shiffrin, 1968; Gazzaniga et al., 2002) that the notion of sensory memory is also taken into account besides the STM. In the AMS context, however, whether such a further distinction is necessary or not may, again, be merely confined within the issue of implementation, as it can be seen that the notion of sensory memory in the structural sense is subsumed under the concept of the STM/working memory module and/or is already implemented within the **sensation** module; for instance, the length of the feature data in each pre-processing unit in Fig. 6.1 may be closely tied to the capacity of sensory memory. (The issue of implementation within the kernel memory concept will also be discussed later in Sect. 8.3.4.)

Although the full account/justifications for the functionality of the STM/ working memory in a cognitive scientific/psychological view point cannot be given in this book, we next consider one of the most influential working memory models describing the "phonological loop" concept, which was originally developed by Baddeley and Hitch (Baddeley and Hitch, 1974), and how such a model can be interpreted within the AMS context.

8.3.1 Interpretation of Baddeley & Hitch's Working Memory Concept in Terms of the AMS

In the psychological study (Baddeley and Hitch, 1974), Baddeley and Hitch proposed the model of working memory which extends the concept of STM such as the one in (Atkinson and Shiffrin, 1968), by introducing the concept of the so-called "phonological loop", with some supportive neuropsychological arguments by the studies of patients with specific brain lesions (for the detail, see e.g. Gazzaniga et al., 2002). Their working memory is divided into three parts, i.e. a central executive mechanism and the two subordinate systems, namely, the phonological loop and visuospatial sketchpad, the latter two of which are controlled by the central executive system. Then, they explained both the forgetting mechanism of STM and the relation between the STM and LTM, e.g. the notion of how the transfer of memory from the STM to LTM can be performed, in terms of their working memory model. As the name "phonological loop" implies, the subordinate system is a mechanism for acoustically (or verbally) coding the information (i.e. sound inputs) in working memory and is considered to perform the coding by subvocally rehearsing the items to be remembered over the short-term. In contrast, the "visuospatial sketchpad" functions separately from (but in parallel to) the phonological loop and performs the coding of the pure visual (or visuospatial) counterpart of the information within the working memory.

Moreover, it is anatomically considered that, apart from the well-known Brodmann's area 40 (Brodmann, 1909), the rehearsal process in the phonological loop involves a region in the left premotor region (area 44), i.e. both the lateral frontal and inferior parietal lobes, whilst for the visuospatial sketchpad the parieto-occipital regions of both the left and right hemispheres of brain are the keys (for a concise review, cf. Gazzaniga et al., 2002)[1].

As in Fig. 5.1 (on page 84), the STM/working memory module has the bi-directional connections with the three modules, i.e. 1) **attention**, 2) **emotion**, and 3) **explicit LTM** module, whilst the **sensation**, **implicit LTM** module, and the two output modules, i.e. both the **primary output** and **perception** (i.e. **secondary output**) modules, are all connected with mono-directional data flows. The latter two represent the feedback inputs to the STM/working memory module. Moreover, the two modules, i.e. 1) **thinking** and 2) **intention** module, are considered to function in parallel.

Hence, it is considered that the model of the aforementioned STM/working memory concept (Atkinson and Shiffrin, 1968; Baddeley and Hitch, 1974; Gazzaniga et al., 2002) is directly relevant to the interactive data processing between the STM/working memory and LTM (and/or the LTM oriented) modules, within the AMS context.

Then, it is considered that the model of working memory proposed by Baddeley and Hitch (Baddeley and Hitch, 1974; Baddeley, 1986) involves the following two data processes:

1) The *data-fusion* of both the auditory and visual sensory data within the STM/working memory module ;
2) The transfer of the outcome within the STM/working memory to the LTM module.

In the AMS context, the two processes in the above can be justified within the interactive data processing between the STM/working memory and LTM modules, as described next.

---

[1]In general AI, it is considered that, although such an anatomical placement for each functionality as described in the above is not always a crucial matter for modelling various cognitive/psychological functionalities, specifying the area/region for a certain function (i.e. the phonological loop/visuospatial sketchpad in the working memory) can greatly facilitate in "understanding" of such function. However, since not only a real brain is a totally complex system but the measurements currently available are limited in the capacity, to elucidate precisely the functionalities, such area/regional specification still remains a hard task. Nevertheless, where appropriate, we consider this sort of anatomical place justifications in this book.

8.3.2 The Interactive Data Processing:
the STM/Working Memory $\longleftrightarrow$ LTM Modules

In the data process 1) above, it is firstly considered that both the auditory and visual sensory data, which are received from the **perception** module and/or recalled from the **LTM** modules (i.e. due to the requests from other associated modules such as **attention** or **emotion**), reside within the STM/working memory module over a certain (short) period of time. Imagine a situation e.g. that the STM/working memory module receives the auditory sensory (encoded) data from the sensation module, which has not yet been stored within a specific area of the LTM, whilst the visual data corresponding to the auditory counterpart have already been stored in advance (by the prior learning process; see Chap. 7) and recalled from the (modality-specific area of) LTM within the STM/working memory. (Thus, the former process represents the data flow; **sensation** → **STM/working memory** module, whereas the latter; **LTM** → **STM/working memory** module)

Then, a reinforcement (or target) signal is given (in a certain manner, i.e. by the interactive processes between the memory modules, as described in the previous chapter) to associate the auditory data received from the sensation module with the visual counterpart via the learning process. In the sequel, this can cause the "data-fusion" of both the auditory and visual data. In terms of the kernel memory, this data-fusion process can be ultimately interpreted as (merely) establishing a connection between one kernel unit with the template vector set to the auditory data and another with the visual counterpart, within the STM/working memory module. For representing this establishment, the principle of SOKM (in Chap. 4), in which the simultaneous activation of the kernel units can eventually lead to the formation of the link weight(s) in between, can be exploited. Hence, it is also said that this process simulates a general notion of learning, e.g. the situation where a child is about to learn/associates the visual part of a new word ("learnt by heart" in advance) with the auditory counter part.

Next, for the data process 2) above, the data transfer, which represents the data flow, i.e. **STM/working memory** → **LTM** module(s), can occur, if (as in the aforementioned phonological loop concept) the outcome of the data-fusion, which can be given in the form of a kernel network consisting of multiple kernel units within the STM/working memory, resides within the STM/working memory for a certain (sufficiently long) period of time. In this regard, it is said that the data transfer, i.e. the STM/working memory → LTM modules, simulates the role of the hippocampus in the neurophysiological context (for a concise review of the studies, see e.g. Gazzaniga et al., 2002).

Therefore, in summary, by examining the two data processes 1) and 2) above, the following three data flows between the three modules, i.e. the STM/working memory, LTM, and the input: sensation modules, can be drawn, as depicted in Fig. 5.1:

- **Sensation ⟶ STM/Working Memory Module**
  Represents the receipt of the (encoded) data from the sensation module; the sensory data will be used for the data-fusion within the STM/working memory module.
- **STM/Working Memory ⟶ LTM Modules**
  Represents the transfer of the transient data or consolidation of the kernel networks (i.e. composed by multiple kernel units and the link weights in between), which have survived after a sufficiently long period of time, within the STM/working memory module to the LTM module(s). In addition, this sort of transfer/consolidation can be occurred intermittently.
- **LTM Modules ⟶ STM/Working Memory Module**
  Represents the memory recall of the data stored within the LTM module(s); as in the first data flow: **sensation ⟶ STM/working memory module**, the recalled data will also be used for the data-fusion within the STM/working memory module, where necessary.

Although the description of the three data flows in the above is limited to the case of the data-fusion where both the auditory and visual data are only considered, within the AMS context, this can be generalised to any combination of the sensory data, without loss of generality.

8.3.3 Perception of the Incoming Sensory Data in Terms of AMS

In AMS, it is considered that, once sensory data are received by the AMS, the perception is (normally) performed via the STM/working memory module; after receiving the sensory data from the **sensation** module, the data are directly transformed into the respective kernel units within the STM/working memory module and also sent to the corresponding modality-specific area of the **implicit LTM** module. Then, the data transfer to the implicit LTM module immediately yields (a series of) the perceptual outputs obtained as the pattern recognition results from the **perception** module (as described in Chap. 6. Hence, in such a case, it can also be seen that the STM/working memory acts as the sensory memory). Eventually, the recognition results are fed back to the STM/working memory module; the perceptual outputs which are given as the feedback inputs to the STM/working memory module may be alternatively represented by the symbolic kernel units (with the kernel function given as (3.11)).

Therefore, performing the perception of the sensory data in terms of AMS involves the following four data flows:

1) **Sensation**   ⟶ **STM/Working Memory**
2) **STM/Working Memory** ⟶ **Implicit LTM**
3) **Implicit LTM**   ⟶ **Perception**
4) **Perception**   ⟶ **STM/Working Memory**

Normally, it is considered that the perception of the incoming data in 1–4) above can be immediately performed. However, how rapidly/correctly the data processing within 1) and 2) can be performed also depends upon the current states of the STM/working memory and the associated modules (i.e. **attention, emotion, intention**, and/or **thinking** module), as described later.

Although the descriptions of the data flows between the STM/working memory and other associated modules, such as attention or emotion, are left to the later chapters, we are now ready to consider modelling the STM/working memory module in terms of the kernel memory, as described in the next subsection.

8.3.4 Representation of the STM/Working Memory Module in Terms of Kernel Memory

Figure 8.1 shows an illustration of the STM/working memory module in terms of the kernel memory representation and the relationship between a total of the nine associated modules, i.e. 1) **attention**, 2) **emotion**, 3,4) both **explicit and implicit LTM**, 5) **intention**, 6,7) both **primary and secondary (perceptual) outputs**, 8) **sensation**, and 9) **thinking** module (also, compare Fig. 8.1 with Fig. 5.1 on page 84).

As in the figure, the STM/working memory module consists of multiple kernel units, as well as the explicit/implicit LTM modules, and is (partially) connected to both the LTM modules, by means of the link weights between the kernel units K_i^S ($i = 1, 2, \ldots, N_S$)[2] and K_j^E and/or K_k^I ($j = 1, 2, \ldots, N_E, k = 1, 2, \ldots, N_I$), where, in each memory module, the number of kernel units is (in practice) assumed to be upper limited, i.e. $N_S \leq N_{S,max}$, $N_E \leq N_{E,max}$, and $N_I \leq N_{I,max}$.

In Fig. 8.1, as indicated by the corresponding data flows, the STM/working memory also receives the feedback inputs from both the primary and secondary (i.e. perceptual) outputs (albeit not explicitly shown for the latter in Fig. 8.1), apart from the sensory inputs; in practice, the STM/working memory module is initially considered as an empty kernel memory space, and, whenever either the incoming data from the sensation module or the feedback inputs from the primary/secondary (i.e. perceptual) output modules are given to the STM/working memory, we may i) create new kernel units one by one or ii) replace some existing ones (i.e. by taking into account the factor N_s).

---

[2]For convenience, in Fig. 8.1, the kernel units with the superscript "S" stands for those within the "STM/working memory", whereas the superscripts "E" and "I" denote respectively the "explicit LTM" and "implicit LTM". In addition, note that, as aforementioned, since here both the sensory memory and STM are treated within a single module in the AMS context, the maximum number of the kernel units $N_{S,max}$ may be set to a relatively large value, by taking into account the large capacity of sensory memory compared to the STM (for this argument, see p.305 of Gazzaniga et al., 2002).

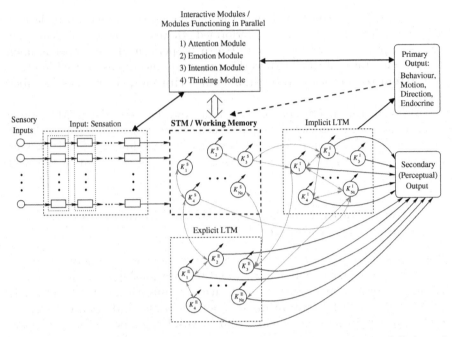

Fig. 8.1. An illustration of the STM/working memory module in terms of the kernel memory, consisting of multiple kernel units, and the relationship between the nine associated modules, i.e. 1) **attention**, 2) **emotion**, 3,4) both **explicit and implicit LTM**, 5) **intention**, 6,7) both **primary and secondary (perceptual) outputs**, 8) **sensation**, and 9) **thinking** module

For both the cases i) and ii), such kernel units are formed, with the template vectors (or matrices) identical to those incoming data/feedback inputs within the STM/working memory module. Then, the data, which are stored in the form of the template vectors within the kernel units so formed, will be immediately sent to the areas corresponding to the respective modality-specific areas of the kernel units within the LTM modules. Thus, in the case of presenting them to the implicit LTM, we may obtain (a series of) the perceptual outputs (e.g. of a particular object(s)) from the secondary output module, which can be given as the cause of the activations of the kernel units within such areas of the implicit LTM module.

For the feedback inputs, it is also possible that they can be (alternatively) represented in terms of symbolic kernel units, instead of exploiting the regular kernel units.

8.3.5 Representation of the Interactive Data Processing Between the STM/Working Memory and Associated Modules

In the later part in Sect. 8.3.2, the three data flows relevant to the STM/ working memory module; i.e. 1) **sensation** $\longrightarrow$ **STM/working memory**; 2) **STM/working memory** $\longrightarrow$ **LTM modules**; and 3) **LTM modules** $\longrightarrow$ **STM/working memory**, were established, by examining Baddeley and Hitch's working memory concept. In this subsection, we consider how these processes can be actually represented within the kernel memory principle.

1) Data flow: Sensation $\longrightarrow$ STM/Working Memory

In Fig. 8.1, the data processing 1) **sensation** $\longrightarrow$ **STM/working memory** is represented by the data flow from the **sensation** module (which consists of a cascade of the pre-processing units, as described in Chap. 6) to the STM/working memory module; the encoded data obtained through a series of the pre-processing units are directly i) given as the input to or ii) used as the respective template vectors to form the kernel units within the STM/working memory. (For the former i), if we consider a Gaussian kernel unit as given by (3.8), the input vector $\mathbf{x}$ corresponds to such encoded data. For either the case i) or ii), we may consider the principle similar to the construction of the SOKM given in Sect. 4.2.4.

2) Data flow: STM/Working Memory $\longrightarrow$ LTM

Then, for representing the data flow 2) **STM/working memory** $\longrightarrow$ **LTM modules**, it is considered that there are the two types of processing involved; i) generation of the perceptual outputs via the LTM modules, due to the activations of the kernel units within the STM/working memory module as aforementioned in the previous subsections, i.e. by the incoming sensory data or thinking process, and ii) the transfer (or transition) of the kernel units from the STM/working memory to the LTM modules (as in the Baddeley and Hitch's working memory described in Sect. 8.3.1).

For ii), a condition must be given to the STM/working memory module; the kernel units swiftly disappear from the STM/working memory module[3], or are replaced by those with different parameter settings, as aforementioned, unless they are transferred to the LTM modules within a certain period of time.

---

[3]In the case of hardware representation, it does not imply that such "disappearance" of the kernel units can actually occur, but rather, the parameters of some kernel units, i.e. the template vectors, link weights, etc, can be reset/become completely different, e.g. when new incoming data arrive at the STM/working memory module.

3) Data flow: LTM ⟶ STM/Working Memory

Thirdly, the data flow 3) **LTM modules** ⟶ **STM/working memory** depicts the recall of the data stored within the LTM modules, due to e.g. the request by the other associated modules.

However, as aforementioned in Sect. 8.2, the third data flow does *not* always imply that the kernel units are actually transferred back (or copied) from the LTM to the STM/working memory module, but, rather, the activated kernel units within the LTM modules are just *monitored* by marking them and then holding the information of the absolute locations, etc, within the auxiliary memory space[4] that may alternatively represent the STM part of the STM/working memory module. In the AMS context, it is also possible to consider that such auxiliary memory can be represented within the **intention** and **thinking** modules, both of which are considered to work in parallel with the STM/working memory module. (We will then return to this issue in Chaps. 9 (Sect. 9.3) and 10 (Sect. 10.4)).

Within a similar context as above, both the two feedback inputs, i.e. the data flow **primary output** ⟶ **STM/working memory** and that **secondary output** ⟶ **STM/working memory**, are depicted (*dashed lines*) in both Figs. 5.1 (on page 84) and 8.1 (i.e. for the former only, as described earlier), since these feedbacks are already represented by the monitoring process of the activations from the kernel units within the LTM modules, the process of which is performed by the STM/working memory module.

8.3.6 Connections Between the Kernel Units within the STM/Working Memory, Explicit LTM, and Implicit LTM Modules

Now, consider a situation where there are multiple kernel units K_i^S ($i = 1, 2, \ldots, N_s$) formed within the STM/working memory, as in Fig. 8.1, and each kernel unit K_i^S is represented in either form depicted in Fig. 3.1 (on page 32) or Fig. 3.2 (on page 37). Then, as illustrated in Fig. 8.1, it is considered that there can be the following five types of the connections between the kernel units (via the link weights):

   i) Connection between K_i^S and K_j^S ($i \neq j$) ;
   ii) Connection between K_i^S and K_j^E or K_k^I ;
   iii) Connection between K_i^E and K_j^E ($i \neq j$) ;
   iv) Connection between K_i^E and K_j^I ;
   v) Connection between K_i^I and K_j^I ($i \neq j$)

The establishment of the connections as in the above can be achieved by e.g. following the Hebbian learning principle as in the SOKM (in Chap. 4);

---

[4]Here, the notion of auxiliary memory is different from that of a kernel unit.

i.e. *"when a pair of kernel units A and B are excited[5] simultaneously and repeatedly (during a certain period of time), a new link weight w_{AB} between the two kernels will be formed, or, if there already exists w_{AB}, the value is increased; otherwise, if such repetitive excitation does not occur for a certain period of time, the value of the link weight w_{AB} is decreased, or such link is eventually removed"*.

In the above, it is also implied that, for all the five connection types, the data-fusion between different modalities can occur, since, within the kernel memory concept, any connections between a pair of kernel units are allowed.

In particular, as discussed in Sect. 8.3.2, the connection type ii) can yield the data-fusion as in Baddeley's working memory concept; if the kernel unit K_i^S is formed using particular auditory sensory data, whereas K_j^I represents the visual counterpart within a specific area of the (implicit) LTM module, and if these two are simultaneously (and repeatedly) excited by the given sensory data, the establishment of the link weight between the two kernel units can be regarded as the data-fusion.

Then, the principle similar to this can be immediately applied to the five connection types in the above. However, for the connection types iii-v), little care must be taken; since the kernels K_j^E and K_k^I reside within the explicit and implicit LTM modules, respectively, they are considered to reside far longer than K_i^S within the STM/working memory module. For instance, by exploiting [**the Link Weight Update Algorithm**], which was given in Sect. 4.2.1 (on page 60), both the decrement ξ_{ij} and increment δ must be set sufficiently smaller than those for i) and ii) above.

8.3.7 Duration of the Existence of the Kernel Units within the STM/Working Memory Module

Next, it is also possible to introduce an extended rule within the STM/working memory module; if there is a kernel unit without having any such connection/being excited for a certain period of time, the kernel unit will be eventually and completely removed from the memory space (or replaced with the one with a totally different configuration). As discussed earlier, whether the removal or replacement is more appropriate is, however, dependent upon the manner of actual implementation within the AMS context.

In respect to the replacement of the kernel units, the structure similar to a last-in-fast-out (LIFO) data stack can be exploited (Hoya, 2004b):

> - If the number of the kernel units $N_s \leq N_{s,max}$ within the STM/working memory, add a new kernel unit in to it;
> - Otherwise, replace the least excited kernel unit with the new one.

For evaluating such excitation, the excitation counter ε attached to each kernel unit and/or the modification of the kernel output by (3.30) can be

---

[5]The excitation of such kernel units can be evaluated by (3.12).

exploited; for instance, if the excitation counter ε_i^S stays below a certain threshold for a certain period of time, the kernel unit K_i^S is replaced/removed from the STM/working memory module, where appropriate.

In Chap. 10, an example of the STM/working memory model to construct an intelligent pattern recognition system will be given, with implementing the aforementioned simple LIFO-like mechanism.

Then, the duration of the existence of the kernel units is quite dependent upon the four associated modules, i.e. **attention**, **emotion**, **intention**, and **thinking**, to be described in the subsequent chapters.

In the following section, we then have a closer look at various LTM modules in the AMS.

8.4 Long-Term Memory Modules

As in Fig. 5.1 (on page 84), there are six long-term memory-oriented modules within the AMS:

1) **Explicit LTM**
2) **Implicit LTM**
3) **Instinct: Innate Structure**
4) **Intuition**
5) **Language**
6) **Semantic Networks/Lexicon**

As shown, all the six modules in the above are (normally) considered to function in parallel without consciousness (i.e. the formation or control of these modules is not consciously performed, given the sensory data. We also consider the general issue of consciousness in Chap. 11).

In this section, we consider only the four LTM-oriented modules, i.e. both the explicit and implicit LTM modules, instinct, and semantic networks/lexicon module, since these are descriptive mainly from the memory aspect. The two remaining modules, i.e. the intuition and language modules, remain to be discussed in later chapters, as they need more justifications apart from the memory perspective.

8.4.1 Division Between Explicit and Implicit LTM

In general cognitive science/psychology, it is thought that LTM can be roughly subdivided into two types, i.e. the explicit and implicit LTM. The former LTM is alternatively called as declarative, whereas the latter is interchangeably referred to as "nondeclarative" memory. This division has been considered, since the memory contents of LTM are found to be either consciously accessible or not (see e.g. Gazzaniga et al., 2002), supported by psychological justifications obtained by studying the cases of amnesic patients, and to date the concept still has widely been acknowledged.

As shown in Fig. 5.1, the explicit LTM module within the AMS has a bi-directional connection with the STM/working memory module, which reflects the notion that only the (conscious) access to the explicit LTM module from the STM/working memory module is allowed, whilst the implicit LTM is connected via a mono-directional link; only the data flow **STM/working memory** $\longrightarrow$ **implicit LTM** module (see also Sect. 8.3.5) is considered, and hence the (conscious) memory retrieval via the STM/working memory from the implicit LTM is not allowed.

In respect of the AMS, the division of the LTM into explicit and implicit counterparts can be reasonable, in that, at some situations, the memory retrieval of a series/chunk of the stored data at a time is necessary, without the data processing via the STM/working memory (that is, without consciousness), in order to make a quick action/response e.g. to external stimuli, whereas any bit of information must be directly (or consciously) accessible via the STM/working memory, where required, e.g. to investigate the surrounding situation strategically (i.e. involving the thinking process) by the currently available (multi-domain) sensory data and the reference to the previously acknowledged/preset data and eventually to take necessary actions (i.e. by accessing then activating some of the kernel units within the implicit LTM (or the procedural memory part) e.g. to invoke the relevant motoric actions).

However, as described later in this chapter, the actual manner in the division of the LTM modules still depends upon the implementation.

8.4.2 Implicit (Nondeclarative) LTM Module

In the cognitive scientific study (Gazzaniga et al., 2002), it is shown that the implicit LTM is subdivided into four memory systems; 1) procedural memory, 2) perceptual representation system (PRS)[6], 3) non-associative learning (i.e. habituation and sensitisation), and 4) classical conditioning.

In the AMS, although the above four memory systems 1-4) can be taken into account within the same framework of the implicit LTM module, it is considered that the last two systems 3) i.e. habituation and sensitisation, and 4) classical conditioning, may also be dealt in conjunction with the **instinct: innate structure** module, since in some situations these two seem to be embedded not only due to the learning by the AMS (or the repetitive exposures of the AMS to the surrounding environment) but also dependent upon the

---

[6]In the AMS context, however, it is considered that the role of the perceptual representation system is not only dependent upon the implicit but also explicit LTM module. This view also agrees with the notion of general cognitive scientific/psychological study of memory; as described earlier, the data processing between the explicit and implicit LTM modules is represented by the connections between the kernels K_i^E and K_j^I in Fig. 8.1 (on page 142), which can justify the psychological argument by Squire (Squire, 1987), in that the priming effects (i.e. due to the PRS) are driven not only perceptually but also conceptually or semantically.

innate structure/instinct, e.g. modelling the situation where in creatures the innate structure of offsprings is inherited from their parents/ancestors; for instance, imagine a situation that an infant can show her/his fear when they look at a picture of dinosaurs, without really experiencing them.

As indicated in Fig. 5.1, the contents stored within the implicit LTM module are not directly (or consciously) accessible from the STM/working memory module, but, oppositely, the data stored in the form of the template vectors of the kernel units within the STM/working memory module are transferred to the implicit LTM module, and, unlike the explicit LTM module, the activations (or excitations) of the kernel units within the implicit LTM are transferred further to *either/both* the **primary output** and/or **secondary (perceptual) output**.

For representing the transfer to the primary output module, some patterns of the activations can directly contribute to e.g. cause a series of motoric actions (i.e. movements) by the body, whilst the latter (partly) represents the activity of the PRS.

Then, within the kernel memory principle, it is considered that, due to the corresponding series of the activations from the kernel units within the implicit LTM module, caused by the data processing amongst the other memory-oriented modules (i.e. the explicit LTM or the STM/working memory module), such actions can be eventually carried out. In other words, the activation of the kernel units within e.g. the explicit LTM module is firstly transferred and caused the activation of those in the implicit LTM module via the link weights established in between. Then, such actions can be performed, if (some of) the kernel units so activated in the implicit LTM module are directly connected, or responsible for e.g. controlling the physical mechanism(s) imitating the real (skeletal) muscles or the PRS.

As stated earlier, such a series of activations, however, cannot be monitored in full detail (or with consciousness) by the STM/working memory module but only via the feedback input(s) given from the primary/secondary (perceptual) output.

8.4.3 Explicit (Declarative) LTM Module

Within the explicit LTM, it is generally considered that there are two types of explicit memory, i.e. episodic and semantic memory, where the former represents the autobiographic memory (i.e. the memory related to specific personal events/experiences), whilst the latter involves the general world knowledge/facts (Tulving, 1972; Gazzaniga et al., 2002), both of which can be retrieved consciously, though such distinction still remains a controversial issue in the psychological study of memory (Squire, 1987).

In the AMS, it may, however, be sometimes useful to separate the semantic counterpart from the regular explicit LTM, where appropriate, since, as shown in Fig. 5.1, the semantic networks/lexicon are more closely oriented with the language module from the structural point of view (to be described later in the following subsection and Chap. 9), and thus the treatment may be differed in the actual design.

Note that, although, as aforementioned, there has been a further distinction between the episodic and semantic memory in the explicit LTM within the general cognitive science/psychology context, there in practice seems no significant difference in terms of the representation by kernel units from the memory point of view. This is since, within the context of AMS, these two types of memory can be described in a single framework of the kernel memory; each memory entity, regardless of episodic and semantic, can be represented by a single kernel unit and/or the associations (or the link weights) formed between multiple kernel units.

8.4.4 Semantic Networks/Lexicon Module

As stated earlier, since the semantic networks/lexicon module is also closely related to the language module, it can be useful in practice to consider that the kernel memory of each entity is rather based upon a symbolic representation. Nevertheless, the kernel memory principle still holds, since the units (or nodes) within the semantic networks/lexicon module have to be connected with the kernel units formed within the other associated modules, and such connections must be weighted (and the values of the weights/manners of connections can also be dynamically varied), e.g. via the learning process of the AMS.

We will return to a further discussion of the semantic networks/lexicon module in Chap. 9, since, as aforementioned, the module is intimately related to the language module. Before proceeding next, however, we review how the three LTM modules, i.e. the explicit LTM, implicit LTM, and semantic networks/lexicon, are mutually related within the AMS context, by examining a simple example of learning a new word by the AMS.

8.4.5 Relationship Between the Explicit LTM, Implicit LTM, and Semantic Networks/Lexicon Modules in Terms of the Kernel Memory

As described earlier and illustrated in Fig. 8.1 (on page 142), each of the three LTM modules, explicit LTM, implicit LTM, and semantic networks/lexicon, can be composed by multiple kernel units in terms of the kernel memory concept.

Then, let us consider a situation where specific sensory data (auditory, say, obtained after the process via the STM/working memory module) are firstly stored in the form of a single kernel unit, with the template vector identical to the feature vector of an utterance of a new word, in a particular modality-dependent area of the LTM. In this manner, a cluster of kernel units will be formed to represent other utterances (or samples) of the same word. As a cause of the **learning** performed by the AMS (in Chap. 7), a kernel network (i.e. represented as a sub-SOKM; see Chap. 4) may be formed within the LTM, which generalises these utterances and can respond to such sound patterns (i.e. to yield the activations from some of the kernel units within the sub-SOKM), when the sensory inputs are given to the AMS. It is thus considered that the activations/excitations of the respective kernel units may eventually contribute to the pattern recognition results within the **secondary output** module, as shown in Fig. 5.1 (on page 84).

From another point of view, it is considered that the formation of the sub-SOKM responsible for the auditory part of the new word is related to both the explicit and implicit LTM modules; for representing the explicit part, the formation is based upon a particular set of the utterances presented to the AMS. In other words, the sub-SOKM so formed stores the auditory information of the new word which is acquired through the learning, or the *exposition* of the AMS to a set of several utterances (i.e. each given sequentially time-wise). Hence, this can represent the notion of *episodic* memory, i.e. one of the constituents of the explicit LTM, as acknowledged in general cognitive science/psychology.

In contrast to the explicit LTM aspect, it is considered that the implicit counterpart is also related to the learning process of the new word; as described in Sect. 8.4.2, the PRS part of the implicit LTM may firstly respond to the fragments (or the respective phonemes) of the new word, instead of the whole sound, and then, after the learning process within the AMS, the sub-SOKM responsible for the new word is formed, e.g. by establishing connections between the kernel units representing the corresponding phonemes within the implicit LTM module (and/or the kernel representing the series of the phonemes within the explicit LTM module. This is then somewhat related to the issue of gnostic cells versus ensemble coding in Sect. 4.6 and the concept formation to be described in the next chapter.).

In a similar fashion (and parallel to) the auditory part, another cluster of kernel units representing, e.g. the spelling (i.e. the visual counterpart) of the new word, will be formed within the explicit/implicit LTM module.

Then, during the course of the further learning process, it is considered that, when the data-fusion of the two modalities within the explicit and/or implicit LTM module, i.e. both the auditory and visual counterparts of the new word, occurs and thereby a (symbolic) kernel unit, which can also transfer the activation(s) for one part (i.e. the kernel units within the sub-SOKMs responsible for either the auditory or visual part) to the other, is formed, this implies the *concept formation* within the semantic networks/lexicon module (to be described in the next chapter).

It is also considered that both the explicit LTM and semantic networks/lexicon modules can be rather represented by multiple symbolic kernel units (as in conventional symbolism) that are mutually connected to the kernel units within the three LTM-oriented modules: explicit LTM, implicit LTM, and semantic nets/lexicon, as shown in Fig. 8.1. Then, the difference between the explicit LTM and semantic networks/lexicon may appear in terms of the connections; for the semantic nets/lexicon module, it is considered that the manner of connection is rather *well-ordered* to describe logically the facts/world knowledge, whereas, for the explicit (or episodic) LTM, the connections are formed, strongly dependent upon, e.g. the manner of the sensory data presentation to the AMS and the internal states at the time of such presentation, and thus is not considered to be always well-ordered.

For the latter (i.e. the episodic part), the connections between the kernel units within the implicit LTM can, therefore, play a more significant role to describe the episodic aspect.

This also implies the possibility of the occurrence of the transition from the explicit (i.e. episodic) LTM to the semantic nets/lexicon, in that such well-ordered structure within the semantic nets/lexicon can be formed through a further learning process (or the repetitive experience) of the facts/world knowledge (i.e. due to the reinforcement, as described in Chap. 7), or, in a more macroscopic sense, due to the reconfiguration of the explicit LTM.

8.4.6 The Notion of Instinct: Innate Structure, Defined as A Built-in/Preset LTM Module

For actually designing/developing the AMS, it seems useful to consider the (rather) static part of LTM, besides the aforementioned three LTM modules; as described earlier, the three LTM modules, i.e. the explicit LTM, implicit LTM, and the semantic networks/lexicon, can be reconfigured dynamically during the learning process (in Chap. 7), whereas the instinct module (rather) remains intact during such process.

Provided that we already have sufficient knowledge/information about the properties of the materials/substances for developing e.g. a robot or humanoid, imagine a situation that we are ready to utilise them for developing such a system. Then, it can be useful/necessary to *preset* the values, representing the constraints or properties of such constituents within some specific area of the LTM of the AMS. This is since such information can be directly/indirectly accessible during the interactive processes amongst the modules within the AMS, be taken into consideration during the action planning (i.e. by the **thinking** module) or exploited for giving the target response (i.e. reinforcement signals) during the learning process (see Chap. 7), and can eventually help to suppress excessive amount of data processes, or, ultimately, prevent serious damage to the system; for instance, such information can be represented by our feeling of pain, e.g. if we try to stretch left arm beyond its length (and is therefore also related to the so-called "body versus mind" (or mind-body) issue).

This is the reason why we take another (rather static) LTM module, i.e. instinct: the innate structure, into consideration. Thus, we treat the notion of "instinct" as a *(mostly) static and parameterised set of values*, which provides the information relevant to the physical nature of the body, within the AMS context. The representation of the **instinct: innate structure** module can, however, still be treated within the kernel memory concept.

As shown in Fig. 5.1, it is considered that the innate structure module can greatly influence the internal states of the AMS, such as those within the

intention or **emotion** module, e.g. via the procedural memory part of the **implicit LTM** module (i.e. in Fig. 5.1, the parallel functionality between **instinct** and **emotion** module is considered, in order to represent the indirect influence), which may cause a significant impact upon the data processing via the **STM/working memory** module amongst the other associated modules and, eventually, dramatic changes in the overall behaviours of the AMS.

In real life, it is commonly acknowledged that there are several types of instinct are considered, viz. hunger, thirst, sexual behaviour, sleepiness, etc, all related to the continuous existence of the life/preservation of the species (for a further discussion, see e.g. Rolls, 1999).

In developing any system of artificial life (i.e. such as a humanoid or robot), however, the design of this module is considered to be (or, at least, *must be*) the most difficult part, though, during the development, it should not be always necessary to simulate every instinctive behaviour that all human-beings share; since such artificial objects are *not* developed as the cause of natural consequences but, rather, *designed* so as to meet the demands of human-beings. (As declared in the Statements in the early part of this book, we should however need to take into account, at least, the Asimov's three principles, (Asimov, 1950) at stage of the actual development.) Besides such ethical issues, to determine such preset values one by one may not be a straightforward task, since the amount of such task can be prohibitively huge.

Therefore, to relax this, determining the boundary between the regular LTM and instinct: innate structure modules and how to store the contents, i.e. to classify the memory contents to be stored into those for the explicit LTM, implicit LTM, semantic networks/lexicon, and the built-in instinct: innate structure module, becomes crucial. Nevertheless, such classification in practice also seems to become harder, depending upon how much degree the system to be developed is complex.

In terms of the memory aspect, the innate structure can be hence regarded as one part of the implicit LTM module (i.e. suggesting the parallel functionality between the instinct: innate structure and implicit LTM module, albeit not shown explicitly in Fig. 5.1) where the memory contents can be less altered and remain almost intact (or slowly varied), during the course of the learning process (in Chap. 7), compared to the regular implicit LTM.

8.4.7 The Relationship Between the Instinct: Innate Structure and Sensation Module

As described in Sect. 8.3, the STM/working memory module plays the central role for the interactive processes between other associated modules within the AMS. Then, within the example of general evolutionary process as shown in Fig. 7.1 (on page 118), it is considered that the error signal is also fed back to the sensation module, in order to perform the self-evolutionary process (or the reinforcement learning). However, it is intuitively considered that this feedback data flow can be ultimately ascribed to the relationship between the sensation and the innate structure module, with the notion that such data flow can occur without consciousness (though this is not explicitly shown in Fig. 5.1).

8.4.8 Hierarchical Representation of the LTM in Terms of Kernel Memory

Here, we consider how the LTM modules can be structured by means of the kernel memory concept in a more practical view point. As described in Chaps. 3 and 4, if the Gaussian response function (given in (3.8)) is chosen, the selection of the factor (or the radius) σ can give a significant impact upon the generalisation capability of the kernel memory, and, as seen in the simulation study of the SOKM in Sects. 4.4 and 4.5, a unique setting of this factor is reasonable to yield a satisfactory generalisation performance. However, similar to the case of the PNN/GRNN as described in Sect. 2.3.4, as the size of the kernel network (or sub-SOKM) responsible for a particular category/class becomes large, the computation time required for both updating the radii values and accessing the kernel units in the reference (or testing; cf. Polikar et al. (2001) and Hoya (2003a)) mode may become problematic.

This is hence crucial for the actual design of the memory/memory-oriented modules within the AMS (especially, this is so, if we consider the amount of data to be stored within the LTM modules).

Now, let us consider how to organise the auditory part of the LTM (albeit here putting aside the issue of the sub-structures described earlier, i.e. the explicit LTM, implicit LTM, semantic networks/lexicon, or the innate structure), in order to perform efficiently the auditory data classification tasks by means of the kernel memory representation.

Fig. 8.2 illustrates an example of the tree-like representation of the LTM responsible for auditory data in terms of the kernel memory. (Note that, not limited to the one in Fig. 8.2, any hierarchy can be possible in terms of the kernel memory concept; e.g. the structure taking into account multiple languages can also be considered.)

In the figure, provided that the auditory data (i.e. stored in the form of the template vector of a single kernel unit) is sent from the STM/working memory module, (*approximate*) classification of the auditory data given is firstly performed, i.e. whether the sound data given from the STM/working memory module corresponds to a human voice, music, noise, and so forth (at Level 1). By the term "approximate", it is meant that the classification within each sub-SOKM is *quickly* performed, instead of using the original data, for the efficiency in the computation of searching the huge memory space of the LTM; for instance, by lowering temporarily the resolution in time and/or frequency wise (e.g. with applying a low-pass filter/resampling), such a quick classification can be achieved, since the size of the data becomes smaller than that of the originally given sound data.

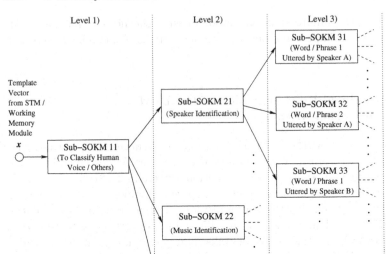

Fig. 8.2. Illustration of an example of the LTM responsible for the auditory data in terms of the hierarchical kernel memory representation – a tree like representation; in this example, at Level 1), the sound data given from the STM/working memory module will be roughly classified, i.e. whether the data corresponds to human voice or another type of source. Then, the subsequent classification of the sound data will be performed; e.g. if the data are found to be a human voice by the pattern matching in Sub-SOKM 11, the speaker identification, then the word/phrase recognition, and so forth. Or, otherwise, whether the data corresponds to music or noise, and so forth ensues. The level of pattern recognition process to be performed depends upon either the current states of the STM/working memory module or the interactive data processing between the associated modules within the AMS

Then, in a similar context, the classification is subsequently performed (e.g. the classification at word level, phonemes, and so forth, due to the transfer of the activations of the kernel units via the link weights across the sub-SOKMs ij; i: the level number, j: the sub-SOKM number), up to the level (of classi- fication) in which some of the kernel units at the corresponding sub-SOKM are found to be activated by the auditory data/required for the further data processing between the other associated modules within the AMS (such as **thinking** or **intention**) via the STM/working memory module.

Moreover, by means of such hierarchical structure, it is also considered that the exhaustive search can be avoided; during the construction/reconfiguration (or the learning) of the LTM, to update i.e. the radii value σ for each Gaussian kernel unit based upon the unique radii setting scheme as in (2.6), the computational load to find the maximal distance can be greatly reduced, since the search will be limited to those within the corresponding sub-SOKM only.

In practice, to construct such hierarchical structure via the learning process of the AMS, we may either pre-determine the hierarchy of the LTM or leave it to the autonomous construction/reconfiguration of the SOKM), depending upon the application. For the latter, however, the hierarchical structure to be formed may be totally different from the one what we expect, e.g. some other form of sub-SOKMs may be constructed to achieve reasonable performance for the auditory data classification tasks. It is then considered that the hierarchical structure within the LTM is quite dependent upon the situation where the AMS is applied (e.g. the manner in the presentation of the auditory data).

8.5 Embodiment of Both the Sensation and LTM Modules – Speech Extraction System Based Upon a Combined Blind Signal Processing and Neural Memory Approach

In this section, we consider the embodiment of the two modules within the AMS described earlier, i.e. the **sensation** and **LTM** modules[7], and the practical application to speech extraction in cocktail party situations.

In a cloud of people, we humans still can easily recognise and then be attentive to the voice of a particular person(s) we know, but how can we simulate this ability by machines? This is referred to as the so-called *cocktail party problem*. In the last decade, with the advancements in the algorithms for blind signal processing, i.e. the independent component analysis (ICA) (see e.g. Cichocki and Amari, 2002), the cocktail party problem has been tackled by a number of researchers (see e.g. Haykin, 2000).

In the recent studies (Barros et al., 2000; Rutkowski et al., 2000; Barros et al., 2002), a variant of subband blind extraction methods based upon the ICA approach has been proposed. As reported, these methods work well to extract the speech component with the highest energy. However, the enhanced speech obtained using these methods can be greatly deteriorated, mainly due to the incomplete reconstruction of the signal. This results from both the scale misadjustment and permutation ambiguity at each subband, which are fundamentally inherent to the ICA. We then incorporate the concept of the LTM modules, which can be represented within the kernel memory principle (in Chap. 3), into the original speech extractor based upon the subband

---

[7]Throughout this section, however, without loss of generality, the aforementioned subdivision of the LTM into the four constituents, i.e. explicit LTM, implicit LTM, semantic networks/lexicon, or instinct: innate structure, is not considered.

ICA, in order to compensate for these two problems and thereby achieve the performance improvement in terms of the speech extraction.

In cocktail party situations, it is assumed that there are M source signals (or M simultaneously uttered voiced speech signals), i.e. $s_i(k)$ ($i = 1, 2, \ldots, M$, at discrete time k). Then, provided that (without loss of generality) we have two microphones (or sensors), the observed signals $\mathbf{u}(k) = [u_1(k), u_2(k)]^T$ can be defined as an under-determined linear convolutive model:

$$\mathbf{u}_i(k) = \sum_{n=-\infty}^{\infty} \mathbf{H}(n)\mathbf{s}(k-n) \tag{8.1}$$

where the source signals (represented in vector form) $\mathbf{s}(k) = [s_1(k), s_2(k), \ldots, s_M(k)]^T$ and $\mathbf{H}$ ($\in \Re^{2 \times M}$) defines the mixture represented by a linear filter operator. In a real acoustic environment, since $\mathbf{H}$ is generally a non-minimum phase low-pass filter (see e.g. Gold and Morgan, 2000), extraction of the original speech signal is very hard.

8.5.1 Speech Extraction Based Upon a Combined Subband ICA and Neural Memory (Hoya et al., 2003c)

Figure 8.3 illustrates the block diagram of the speech extraction system based upon the combined subband ICA and neural memory (Hoya et al., 2003c).

In Fig. 8.3, we firstly obtain the two observations $u_1(k)$ and $u_2(k)$ (both can be given in (8.1)) from the two microphones, and the batch of the data, i.e. $\mathbf{u}_i(k) = [u_i(k), u_i(k-1), \ldots, u_i(k-L+1)]^T$ (L: the analysis window length), is transferred to the respective subband decomposition units 1 to N. After the subband decomposition (Barros et al., 2002), two types of subband signals are obtained: 1) the subband signals $\mathbf{B}_j$ ($j = 1, 2, \ldots, N$, for convenience, we here omit the time instance k.) and 2) amplitude envelope signals $\mathbf{C}_j$. For both the signals $\mathbf{B}_j$ and $\mathbf{C}_j$, we then perform ICA to separate the components. Next, we select either the signal component $\mathbf{l}_{j1}$ or $\mathbf{l}_{j2}$ and adjust the amplitude of the signal at each subband by multiplying the factor d_j, based upon the pattern recognition results, i.e. by performing twice the pattern recognition with presenting (one by one) the signal $\mathbf{r}_{j1}$ and $\mathbf{r}_{j2}$ (both of which are obtained from the j-th ICA unit) to the neural memory unit at the j-th subband (i.e. denoted NM j in Fig. 8.3).

Eventually, the extracted (or reconstructed) target speech $\hat{\mathbf{s}}_1$ (suppose that we want to extract the first source; given the batch data $\mathbf{u}_i$) can be obtained as

$$\hat{\mathbf{s}}_1 = \sum_{j=1}^{N} d_j \mathbf{l}_{jp} \tag{8.2}$$

where $p = 1$ or 2 and is determined by the reference to the neural memory.

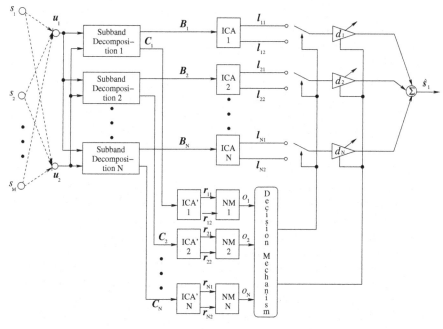

Fig. 8.3. Block diagram of the speech extractor based upon a combined subband ICA and neural memory; the two observations $\mathbf{u}_1$ and $\mathbf{u}_2$ are firstly decomposed into N subband signals $\mathbf{B}_j$ $(j = 1, 2, \ldots, N)$ and converted into the respective amplitude envelopes $\mathbf{C}_j$, by applying the subband decomposition mechanism (Barros et al., 2002). Second, the separated components obtained from the ICA of the amplitude envelopes $\mathbf{C}_j$ are used to identify the signal of interest by the neural memory (NM j). Then, by taking the statistics of the pattern matching obtained from NM j, the subband signal of interest at each subband (i.e. either $\mathbf{l}_{j1}$ or $\mathbf{l}_{j2}$) is determined and thereby the scaling adjustment factors d_j corresponding to the signal of interest are extracted from the auxiliary memory. Finally, the signal of interest will be reconstructed by applying (8.2)

As described in Sect. 6.2, the subband decomposition within the speech extractor can be regarded as one of the fundamental parts within the **sensation** module of the AMS, whereas the neural memory units correspond to the **LTM** modules. Next, we focus upon the subband ICA mechanism and neural memory in more detail.

The Subband ICA Mechanism

In subband blind extraction approaches, it is considered that the effectiveness resides in the property that narrow band signals are less prone to the convolutive effects in comparison with the original fullband signal (Barros et al., 2002). The subband coding scheme (Barros et al., 2002) used in Fig. 8.3 was then developed by exploiting the harmonicity of the voiced sounds, as in some models of computational auditory scene analysis (Weintraub, 1985).

In summary, the coding mechanism (Barros et al., 2002) involves the following two steps:

Step 1) Extraction of the fundamental frequencies, obtained by applying both the wavelet (see e.g. Mallat, 1999) and Hilbert transforms (see e.g. Proakis and Manolakis, 1992).

Step 2) Processing the signal with the bank of adaptive bandpass filters centered at the fundamental frequency f_0, given in Step 1), and its harmonics.

As stated earlier, in the scheme shown in Fig. 8.3, both the two-channel subband signals $\mathbf{B}_j$ ($\Re^{L \times 2}$, $j = 1, 2, \cdots, N$, denoted in block form) and (instantaneous) amplitude envelope signals $\mathbf{C}_j$ ($\Re^{L' \times 2}$), which can be obtained as the intermediary signals in Step 2) above, are exploited.

Then, for the separation of both $\mathbf{B}_j$ and $\mathbf{C}_j$, the ICA algorithm such as second-order blind extraction (SOBI) algorithm (Belouchrani et al., 1993) can be exploited. In brief, the SOBI algorithm is based upon joint diagonalisation of correlation matrices and known to be robust for nonstationary signals such as speech. (In a noisy environment, the robust form of the SOBI (Cichocki and Amari, 2002) can be alternatively exploited.)

As in Fig. 8.3, the SOBI algorithm is independently applied to the separation of $\mathbf{B}_j$ (i.e. ICA j) and $\mathbf{C}_j$ (i.e. ICA' j). Then, it is considered that, due to the statistical invariance between the subband signal $\mathbf{B}_j$ and the amplitude envelope $\mathbf{C}_j$, there is no permutation problem between the respective separated signals $\mathbf{l}_{jp}$ and $\mathbf{r}_{jp}$ ($p = 1, 2$), i.e. both $\mathbf{l}_{jp}$ and $\mathbf{r}_{jp}$ correspond to the same source signal. Although the rigorous and theoretical justification is still under investigation, this was empirically confirmed by the preliminary simulation study.

The Neural Memory

For representing the neural memory, the probabilistic neural networks (PNNs, as described in Sect. 2.3), which can be subsumed within the kernel memory principle (in Sect. 3.2.3), are utilised. (Not to mention, the hierarchical kernel memory representation in Sect. 8.4.8 can be alternatively exploited to construct the neural memory part of the blind speech extractor.)

As aforementioned, the role of the neural memory is to determine:

1) Which subband signal $\mathbf{l}_{jp}$ should be used ;
2) The scale adjustment factors d_j for the re-construction of the speech $\hat{\mathbf{s}}_1$

and therefore, the neural memory part must be constructed (or trained), before performing the blind speech extraction.

Input Vector to the PNN

Then, the pattern recognition by the reference to the neural memory is performed using the amplitude envelope signals C_j, instead of the subband signals B_j, since, for one reason, the size of the matrix C_j can be smaller than that of B_j; the length L' is less than that of each column vector in B_j (i.e. $L' \leq L$; the length L' is proportion to $1/N$, N: the number of subbands), and, for the other, sufficient information about the original signal for the speech extraction is retained in C_j. Thereby, the pattern recognition process can be more efficiently performed than that based upon the subband signals B_j.

During the construction phase of the neural memory, a total of N new RBF units per pattern data will be concurrently created, within the respective PNNs, with each C_j being stored as the centroid vector/matrix of the corresponding RBF. Then, the class ID represents the ID of the (target) speech signal. In addition to this, we also store the values obtained by calculating the standard deviation of B_j (i.e. from noise-free speech) as the scale adjustment factor d_j in the auxiliary memory[8], during the construction phase.

For the reference to/construction of the neural memory, there is, however, one thing that we need to take into account: since the length of the amplitude r_{jp} (and the size of the centroid vectors/matrices so formed) may be varied from one utterance to the other, the input vector to the PNN must be normalised not only in amplitude but also time wise; to adjust the length of the input vector, we can generally consider applying a simple resampling mechanism with both anti-aliasing low-pass filters and zero-padding, where necessary (see e.g. Proakis and Manolakis, 1992).

i) Determination of the Subband Signals for the Reconstruction of Speech Signal of Interest

For the purpose i) above, the determination is thus performed based upon the pattern recognition (or signal identification) result of each blindly separated (and normalised) amplitude envelope r_{jp} given to the neural memory; at the j-th subband, the two separated components r_{j1} and r_{j2}, both of which are obtained from the separation of the amplitude envelope C_j by the j-th ICA unit, are individually given as the input vectors of the PNN and yield the respective pattern recognition results. Then, either of the two components l_{jp} is chosen for the reconstruction of the target speech $\hat{s}_1$, based upon the pattern recognition result; i.e. if r_{j1} is recognised as the amplitude envelope signal of the target speech, the component l_{j1} is chosen for the reconstruction, otherwise l_{j2}. In this manner, we obtain a total of N pattern recognition results for all the N subbands.

Alternatively, we can introduce a simple scoring scheme (Hoya et al., 2003c): if Channel p ($p = 1$ or 2) is more often recognised as the component

---

[8]Here, the notion of auxiliary memory is different from that of a kernel unit.

of the target speech than the other for all the N subbands, regard Channel p as the component corresponding to the target speech and choose uniquely the subband signals l_{jp} ($\forall j$) for the reconstruction. Although the theoretical justification is yet to be given, this simple scheme also works satisfactorily (Hoya et al., 2003c).

Nevertheless, the permutation problem inherent to ICA can be solved by the pattern recognition based scheme in the above.

ii) Scale Adjustment in Each Subband for the Reconstruction

For the second purpose ii), since we know which subband signal should be chosen for the reconstruction of the target speech at each subband in i) above, we simply recall a set of the corresponding scale adjustment values d_j ($j = 1, 2, \ldots, N$) from the auxiliary memory (i.e. stored during the construction phase of the neural memory, together with the amplitude envelope). Finally, we obtain the reconstructed speech by (8.2) by applying these values.

Simulation Examples

We here consider some simulation examples by applying the speech extraction scheme shown in Fig. 8.3. In the simulation example, we consider the two types of two-channel mixture; 1) the instantaneous and 2) (time) delayed mixture. For the delayed mixture 2), we assume the situation where there is only one (or two) dominant speech signal(s) with no time delay and the other background signals with delays and less amplitudes than the dominant speech.

For the simulation examples given here, the number of subbands $N = 64$, and the speech data set consisting of a total of 16 speech utterances spoken in both Portuguese (i.e. a total of $3 \times 4 = 12$ utterances) and Polish (i.e. $4 \times 1 = 4$ utterances), recorded at 8(kHz) sampling rate, was used; the Portuguese utterances were the three digits /NOVE/, /OITO/, and /QUATRO/, each uttered four times by a male and two native female speakers (i.e. each of the three speaker uttered the corresponding digit), whilst the four Polish utterances were /JEDEN/, /ANONIM/, /NAZYWAM SIE/, and /KOWAL-SKI/, each uttered (once) by two male and female native speakers. Then, in the simulation, the task was to extract only the female utterance, i.e. the digit uttered in Portuguese /NOVE/.

For the neural memory, amongst a total of the 12 utterances in Portuguese, the three utterances (out of four) for each digit, i.e. /NOVE/, /OITO/, and /QUATRO/, (hence. a total of nine utterances), were used for constructing (or training) the respective PNNs and auxiliary memory, and the remaining one for each digit was used for generating the mixtures (i.e. for the utility in the reference/testing mode). This is hence to simulate the LTM responsible for a single language (i.e. the Portuguese) by the neural memory. Then, the length of both the input and centroid vectors within the PNNs was fixed to

64, after the aforementioned normalisation (i.e. in both amplitude and time-wise).

To obtain a single set of the scale adjustment factors d_j for the auxiliary memory, the averaged values over the three utterances data for each digit (i.e. used for the training data set) were computed and stored as the respective values of d_j, since, during the simulation, it was confirmed that each Portuguese digit was uttered by a single speaker with almost no variance in the pronunciation during the recording session.

Simulation Example 1) – the Instantaneous Mixture Case

Figure 8.4 shows the simulation result of the instantaneous mixture, where there are assumed the five simultaneous voices (i.e. mingled with the two languages: the three Portuguese digits s_1-s_3 and two Polish utterances s_4 (/JADEN/) and s_5 (/ANONIM/)):

$$u_1(k) = 0.8s_1(k) + 0.4s_2(k) + 0.2s_3(k) + 0.1s_4(k) + 0.3s_5(k) \; ;$$
$$u_2(k) = 0.6s_1(k) + 0.8s_2(k) + 0.3s_3(k) + 0.2s_4(k) + 0.1s_5(k) \; .$$

In the mixture model above, it was also assumed that two Portuguese speakers (i.e. s_1 and s_2) are dominant and the remaining three are the background signals.

In Fig. 8.4, the performance is compared with the three different approaches: 1) the original subband approach (Barros et al., 2002) with the SOBI algorithm (Belouchrani et al., 1993) (i.e. the signal z_1), 2) the fullband SOBI scheme (i.e. applying the SOBI algorithm directly to the two observations u_1 and u_2 without the subband decomposition), and 3) the combined subband SOBI with the neural memory approach.

In the simulation, for 3), Channel 1 was recognised as the voice corresponding to the Portuguese digit /NOVE/ (i.e. applying the aforementioned scoring scheme in Hoya et al. (2003c)) and the speech signal $\hat{s}_1(k)$ was reconstructed, according to this recognition result.

Simulation Example 2) – the Delayed Mixture Case

For the delayed mixture case, the following mixture model consisting of the three simultaneous voices (i.e. the three Portuguese digits: s_1-s_3) was assumed:

$$u_1(k) = 0.8s_1(k) + 0.2s_2(k - 120) + 0.3s_3(k - 180) \; ;$$
$$u_2(k) = 0.6s_1(k) + 0.2s_2(k - 125) + 0.3s_3(k - 190) \; .$$

Figure 8.5 then shows the simulation result. For the combined approach 3), similar to the instantaneous mixture case, Channel 1 was recognised as the corresponding target speech /NOVE/, for the delayed mixture case.

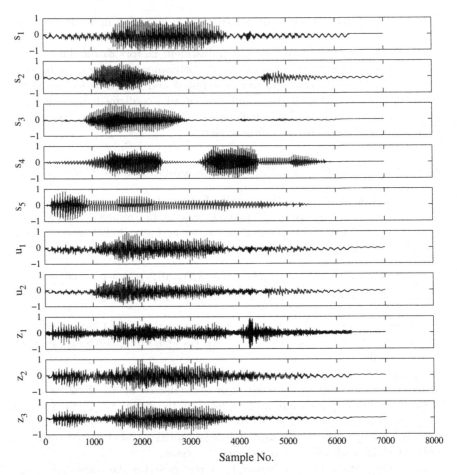

Fig. 8.4. A simulation example of blind speech extraction – the instantaneous mixture case where five simultaneous voices are assumed; i.e. mingled with the two languages: the three Portuguese digits s_1-s_3 and two Polish utterances s_4 (/JADEN/) and s_5 (/ANONIM/); u_k ($k = 1, 2$): the two instantaneous mixtures, the extracted target speech z_1: original subband SOBI, z_2: fullband SOBI, and z_3: the combined subband SOBI with the neural memory approach

Objective Measurement: the Energy in Difference

In both Figs. 8.4 and 8.5, it is observed that the extraction (or separation) performance of the combined approach 3) (i.e. z_3) is better than the other two, i.e. 1) the original subband SOBI (z_1) and 2) fullband SOBI (z_2). However, during the (informal) listening tests, it was sometimes not noticeable that the combined approach 3) yields consistently better performance than the fullband SOBI approach, especially when the number of simultaneous

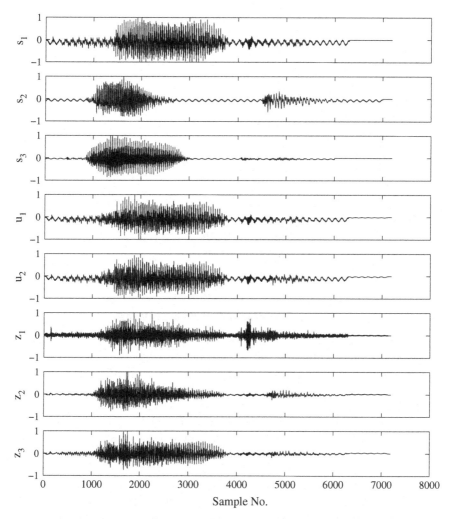

Fig. 8.5. A simulation example of blind speech extraction – the delayed mixture case where three simultaneous voices of a single language are assumed: the three Portuguese digits s_1-s_3; u_k ($k = 1, 2$): the two time delayed mixtures, the extracted target speech z_1: original subband SOBI, z_2: fullband SOBI, and z_3: the combined subband SOBI with the neural memory approach

speech signals was increased for both the instantaneous and delayed mixture cases. Then, in order to confirm this observation and evaluate the performance objectively, we consider the *energy in difference*, E_{diff}:

$$E_{diff} = \frac{1}{q} \sum_{i=1}^{q} \left(\frac{\hat{\mathbf{s}}_1}{\text{std}(\hat{\mathbf{s}}_1)} - \frac{\mathbf{s}_i}{\text{std}(\mathbf{s}_i)} \right)^2 \qquad (8.3)$$

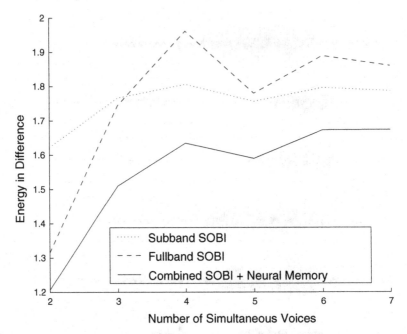

Fig. 8.6. Comparison of E_{diff} – the instantaneous mixture case; as a function of the number of the simultaneous voices in the mixture from two to seven

where q is the number of simultaneous speech signals in the mixture (i.e. known *a priori*), $\mathrm{std}(\cdot)$ denotes the standard deviation, $\hat{s}_1$ is the extracted signal corresponding to the target speech, and s_i are the speech signals. Thus, the quality of the extraction performance improves as E_{diff} becomes small.

Figures 8.6 and 8.7 show respectively the comparisons of E_{diff} (defined in (8.3)) for both the instantaneous and delayed mixture cases, with varying the number of simultaneous voices from two to seven within the mixture. In both the cases, it is evident that the performance with the combined subband SOBI with the neural memory approach 3) (i.e. depicted in *solid lines*) is almost consistently superior to the other two. This is particularly remarkable for the instantaneous mixture case.

8.5.2 Extension to Convolutive Mixtures (Ding et al., 2004)

In the study (Ding et al., 2004), the concept of the neural memory was also incorporated to another blind speech/sound extraction scheme for convolutive mixtures based upon complex ICA in the time-frequency domain, which can then replace the subband SOBI part in Fig. 8.3; in the approach, the subband coding is represented by the ordinary fast Fourier transform (FFT, see e.g. Proakis and Manolakis (1992)), instead of applying a bank of bandpass filters,

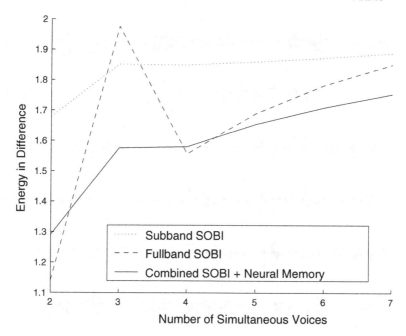

Fig. 8.7. Comparison of E_{diff} – the delayed mixture case; as a function of the number of the simultaneous voices in the mixture from two to seven

and then the complex signals are processed via the respective complex ICA units, whereas the neural memory is, again, realised by the PNNs.

In the simulation, unlike the previous two examples, the task was to extract two out of the three sound sources, i.e. s_1: a female's solo singing voice, s_2: a solo-violin, s_3: a speech utterance in English by a male speaker (i.e. all the three sources were sampled at 8(kHz)), from the two convolutive mixtures u_1 and u_2 with the room impulse responses with the length 64:

$$u_1(k) = \sum_{p=1}^{64} a_{11}(p)s_1(k-p+1) + \sum_{p=1}^{64} a_{12}(p)s_2(k-p+1) \, ,$$

$$u_2(k) = \sum_{p=1}^{64} a_{21}(p)s_1(k-p+1) + \sum_{p=1}^{64} a_{22}(p)s_2(k-p+1) \qquad (8.4)$$

where the impulse responses a_{ij} $(i, j = 1, 2)$ were originally the four distinct paths measured for the two different sound emission positions (i.e. $i = 1, 2$) and two microphones (i.e. $j = 1, 2$), sampled at 24(kHz) in a room of 3.5(m) × 7(m) × 3(m), and down-sampled to a total of 64 coefficients. In the room, each sound emission point was situated with a distance of 0.8(m) in between and away from the two microphones at 1.2(m). Then, the two microphones were located with a distance of 0.58(m) in between.

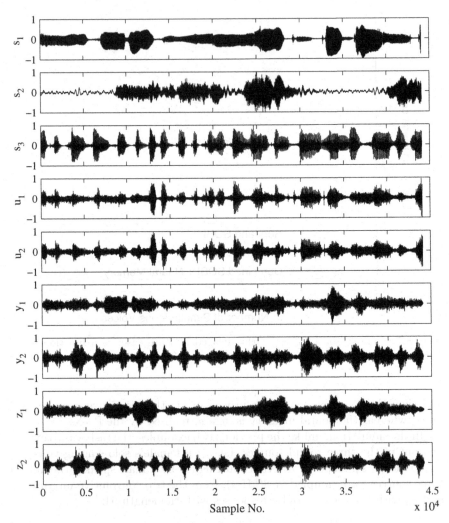

Fig. 8.8. A simulation example of the combined complex ICA in the time-frequency domain and the neural memory for blind sound extraction – the convolutive mixture case (i.e. convolved with real room impulse responses) where a single voice and two simultaneous music sources are assumed: s_1: a female's solo singing voice, s_2: a solo-violin, s_3: a speech utterance in English by a male speaker, u_k ($k = 1, 2$): the two convolutive mixtures, the two sounds extracted y_k: conventional approach in (Murata et al., 2001), and z_k: the combined complex ICA with the neural memory approach

Figure 8.8 shows a simulation example of the combined complex ICA in the time-frequency domain with the neural memory approach. In the figure, the two sounds extracted were the female's vocal singing s_1 (i.e. obtained from Channel 1) and the speech utterance s_3 (i.e. Channel 2). As shown in

Fig. 8.8, the performance of the combined complex ICA with the neural memory approach (i.e. z_θ, $\theta = 1, 2$) was compared to that of the conventional blind speech separation scheme (Murata et al., 2001) (i.e. the plot shown by y_θ).

As confirmed by the listening tests, it is shown that the combined complex ICA with the neural memory approach yields a better performance, in comparison with the conventional approach; in Fig. 8.8, it is remarkable e.g. by examining the segments of y_1 and z_1 between the sample numbers at around 15000 and 30000.

8.5.3 A Further Consideration of the Blind Speech Extraction Model

As described, the neural memory within the blind speech extraction model as shown in Fig. 8.3 can compensate for the problems of permutation and scaling ambiguity, both of which are inherent to ICA. In the AMS context, the subband ICA can be viewed as one of the pre-processing units within the sensory module to perform the speech extraction/separation, whilst the neural memory realised by the PNNs represents the LTM.

Although a great number of approaches have been developed based upon the blind signal processing techniques such as ICA (see e.g. Cichocki and Amari, 2002) to solve the cocktail party problems, the study by Sagi et al. (Sagi et al., 2001) treats this problem rather differently, i.e. within the context similar to pattern recognition/identification. In the study, they exploited sparse binary associative memories (Hecht-Nielsen, 1998) (or, what they call, "cortronic" neural networks), which simulate the functionality of the cerebral cortex and are trained by a Hebbian type learning algorithm (albeit different from the one used in Chap. 4), and their model requires only a single microphone, unlike most of the ICA approaches.

Similar to the pattern recognition context as implied in (Sagi et al., 2001), another model of (blind) speech extraction can be considered by exploiting the concept of learning (in Chap. 7) and the LTM modules within the AMS context; suppose that, within a certain area(s) of the LTM modules, some kernel units are already formed and can be excited by the (fragments of) voice uttered by a specific person, these kernel units can be activated directly/indirectly (i.e. via the link weight(s) from the other connected kernel units), due to the auditory data arrived at the STM/working memory module. Then, as the cause of the interactive processes between the associated modules within the AMS, the state(s) within the **attention** module (to be described in Chap. 10) is varied, the AMS may become attentive to the particular set of auditory incoming data which corresponds to that specific person. Thus, this approach is, in a wider sense, also referred to as the auditory data processing in the cocktail party situations. We will extend this principle to a part of the language processing mechanism within AMS in the next chapter.

8.6 Chapter Summary

This chapter has been devoted to the five memory/memory-oriented modules within the AMS, i.e. 1,2) both the **explicit and implicit LTM**, 3) **STM/working memory**, 4) **semantic networks/lexicon**, and the 5) **instinct** modules, and their mutual relationship, which gives a basis for describing various data processes within the AMS.

As described in Sect. 8.3, the STM/working memory module plays a central part for the interactive data processing between the other associated modules within the AMS.

Within the AMS context, the semantic networks/lexicon module is considered as the part of explicit (declarative) LTM and more closely related to the **language** module than the regular (or episodic) explicit LTM. It is described that, although this notion agrees with the general cognitive scientific/psychological point of view (see e.g. Squire, 1987; Gazzaniga et al., 2002), the division between the explicit LTM and semantic networks/lexicon depends upon the actual implementation within the kernel memory context. In a similar context, the instinct: innate structure module consists of a set of the preset values (or those slowly varying, represented within the kernel memory principle) representing the constraints/properties of the constituents of the system and thus can be regarded as a rather static part of the implicit LTM. However, as described, the division between the instinct and implicit LTM module is, again, dependent upon the implementation.

In cognitive science-oriented studies (for a concise review, see Gazzaniga et al., 2002), whilst it is considered that the hippocampus plays a significant role for the data transfer from the STM/working memory to LTM (Baddeley and Hitch, 1974; Baddeley, 1986) (as described in Sect. 8.3.1), it is thought that the medial temporal lobe/prefrontal cortex corresponds to the explicit (i.e. both the episodic and semantic parts) LTM, whereas, the three areas, i.e. 1) the basal ganglia and cerebellum, 2) perceptual and association neocortex, and 3) skeletal muscle, are the respective candidates for the procedural memory, PRS, and classical conditioning (see e.g. p.349 of Gazzaniga et al., 2002) within the implicit LTM. Although it is considered that this sort of anatomical place adjustment is not crucial, it can give further insights for the division of the memory/memory-oriented modules within the AMS at the stage of the actual implementation.

9

Language and Thinking Modules

9.1 Perspective

In this chapter, we focus upon the two modules which are closely tied to the concept of "action planning", i.e. the 1) **language** and 2) **thinking** modules.

In contrast to the other modules within the AMS, the two modules will be treated rather differently, in that both the language and thinking modules are considered as the built-in mechanisms/the modules which consist only of a set of rules and manage the data processing between the associated modules.

For the former, in terms of the modularity principle of mind, whether the language aspect of mental activities should be dealt within a single module or a monolithic general-purpose cognitive system has long been a matter of debate (Wilson and Keil, 1999). Related to the modularity of language, the study by Broca performed in 1861 indicates that the third frontal gyrus (now well-known as "Broca's area") of the language dominant hemisphere (i.e. the left hemisphere of the brain for right-handed individuals) as an important language area (Wilson and Keil, 1999). The postulate was later (at least, partially) supported by the study of working memory using modern neuroimaging techniques (Smith and Jonides, 1997; Wilson and Keil, 1999), though the overall picture of language representation is still far from clear, and the issues today are focused not upon identifying the specific areas of brain that are responsible for language but rather how the areas of language processing are distributed and organised within the brain (Wilson and Keil, 1999).

Nevertheless, as we will see next, the language module within the AMS context is regarded as a mechanism that consists of a set of grammatical rules and functions as a vehicle for the thinking process performed by the thinking module (Sakai, 2002). On the other hand, within the AMS context, the latter module can be regarded as a mechanism that mainly performs the memory search amongst the **LTM** and LTM-oriented modules and the data processing with the associated modules such as the **STM/working memory** and **intention** modules.

Tetsuya Hoya: *Artificial Mind System – Kernel Memory Approach*, Studies in Computational Intelligence (SCI) **1**, 169–187 (2005)
www.springerlink.com   © Springer-Verlag Berlin Heidelberg 2005

As in Fig. 5.1 (on page 84), it is then considered that both the modules of language and thinking work in parallel, and, as discussed in the previous chapter, the two modules are closely tied to the concept of memory within the AMS context; it is considered that the language module is also closely oriented with the semantic networks/lexicon module and hence the explicit/implicit LTM modules, whilst the thinking module also functions in parallel with the STM/working memory module.

9.2 Language Module

Although the concept of language and how to deal with the notion for the description of mind may vary from one discipline to another (see also Sakai, 2002), within the AMS context, the module of language is defined not as a built-in and completely fixed device without allowing any changes in the structure but as a dynamically reconfigured learning mechanism (cf. the link between the **innate structure** and language module shown in Fig. 5.1 and the description in Sect. 8.4.6), consisting of a set of grammatical rules, and functions as a vehicle for the thinking process performed by the **thinking** module (Sakai, 2002) (thus, the parallel functionality between the language and thinking module is considered within the AMS context, as indicated by the link in between in Fig. 5.1). In respect to the innateness in this wider sense, the notion of the language module within the AMS context coincides with the general concept proposed by Chomsky (Chomsky, 1957; Sakai, 2002), though some principle within his concept, e.g. the universal language theory, has raised considerably certain controversial issues amongst various disciplines (for a concise review, see e.g. Wilson and Keil, 1999)[1]. In contrast, in some recent studies, it is, however, considered that Chomsky's deep thought about language has often been misinterpreted (e.g. Taylor, 1995; Kawato et al., 2000; Sakai, 2002).

Nevertheless, we here do not dig further into such disputes, i.e. those which are related to the justification/validation of Chomsky's concept, but consider, only from the structural point view and for the purpose of designing the AMS, that the language module itself is not completely fixed, but rather, the language module can also be dynamically evolved in nature during the learning process. (For the detail, see Sakai (2002)).

From the linguistic view (Sakai, 2002), it is also considered that the acquisition of the grammatical structure[2] in a language is related to the role of

---

[1]The issue of how to divide actually the language module into the mechanism that is considered to be dependent upon the innate structure and reconfigurable counterpart is beyond the scope of this book. Nevertheless, within the AMS context, it seems appropriate to consider that the language module has the relationship with the instinct: innate structure module (as indicated by the link in between in Fig. 5.1).

[2]With respect to the acquisition of the grammatical structure (and implementation within the AMS), the research is still open (Sakai, 2002); i.e. more studies in

the procedural memory within the implicit LTM, whilst the explicit LTM (or the declarative memory) corresponds to the learning of "meaning" (or the semantic sense of LTM). (For the latter, the notion then agrees with the general principle in cognitive science/psychology, as described in Chap. 8).

More specifically, the learning mechanism represented by the language module within the kernel memory principle is also responsible for the reconfiguration of the **semantic networks/lexicon** module, and thus for the formation of the link weights between the kernel units within the other LTM/LTM-oriented modules and those within the semantic networks/lexicon (as described in the previous chapter) module, so that e.g. the concept formation (to be described later in this section) is performed. However, the manner of such reconfiguration/formation of the link weights can be strongly dependent upon the innate structure of the AMS. (For the general principle of the learning within the AMS context, also refer back to Chap. 7.) In the sense of the innateness, it is said that Chomsky's idea of language acquisition device (LAD) (Chomsky, 1957) can moderately or partially agree with the learning principle of the language module within the kernel memory context.

We next consider how the semantic networks/lexicon module can be actually designed in terms of the kernel memory principle, by examining through an example of the kernel memory representation.

9.2.1 An Example of Kernel Memory Representation – the Lemma and Lexeme Levels of the Semantic Networks/Lexicon Module

In the study by Levelt (Levelt, 1989; Gazzaniga et al., 2002), it is thought that the organisation of the mental lexicon in humans can be represented by a hierarchical structure with three different levels, i.e. the 1) conceptual, 2) lemma, and 3) lexeme (sound) levels.

In contrast, the kernel memory representation of the mental lexicon can be considered to consist essentially of only two levels, i.e. the 1) conceptual (lemma) and 2) lexeme levels, as illustrated in Fig. 9.1, though the underlying principle fundamentally follows that by Levelt (Levelt, 1989; Gazzaniga et al., 2002).

In terms of the kernel memory representation, it is considered that both the lemma and lexeme levels are composed of multiple clusters of the kernel units, as shown in Fig. 9.1.

In Fig. 9.1, without loss of generality, only two modalities, i.e. auditory and visual, are considered at the lexeme level. As shown in the figure, for the visual modality of a single language (i.e. English)[3], three types of the clusters are considered; the clusters of kernel units representing i) words in

---

developmental psychology as found in (Hirsh-Pasek and Golinkoff, 1996) are considered to be beneficial.

[3]In terms of the kernel memory principle, the extension to multiple languages is straightforward.

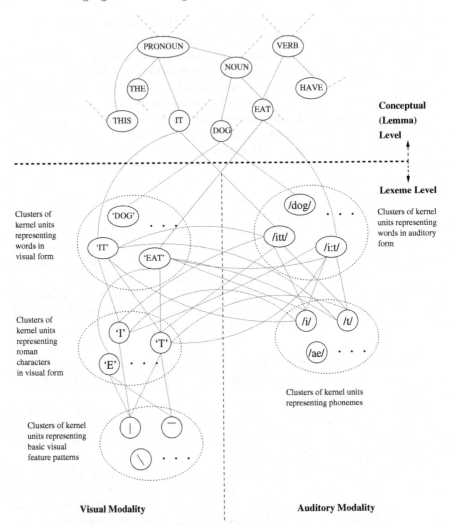

Fig. 9.1. An illustration of the mental lexicon in terms of the kernel memory representation – the fragment of a lexical network can be represented by a hierarchical structure consisting of only two levels: the 1) conceptual/lemma and 2) lexeme levels. Then, each cluster of the kernel units at the lexeme level is responsible for representing a particular lexeme of the lemma and contains multiple kernel units to generalise it. (Note that, without loss of generality, no specific directional flows between the kernel units are considered in this figure)

visual form (i.e. image patterns), ii) Roman characters, which constitute the words in i), and iii) basic visual feature patterns, such as segments, curves, etc, whereas the auditory counterpart contains the two types of the clusters, i.e. those representing iv) words (i.e. sound patterns) and v) phonemes. (Remember that, as described in Chaps. 3 and 4, such cross-modality link weight connections between the respective kernel units are allowed within the kernel memory concept, unlike the conventional ANN approaches.)

For the cluster iii), the well-known neurophysiological study of the cells in the primary visual cortex by Hubel and Wiesel (Hubel and Wiesel, 1977) also suggests this sort of organisation. Then, each cluster in i)-v)[4] is responsible for representing a particular lexeme relevant to the lemma and contains multiple kernel units that generalise it and, in practice, can be formed within the SOKM principle (in Chap. 4).

Figure 9.2 shows an example of the cluster of kernel units representing the sound pattern /iːt/ (/EAT/). (Note that, as defined in Sect. 3.3.1, in both Figs. 9.1 and 9.2, the connections in *grey lines* represent the link weight connections between pairs of the kernel units, whereas those in *black lines* denote the regular inputs to the kernel units, i.e. the data transferred from the STM/working memory module as described in Chap. 8.)

In the figure, it is considered that each kernel unit, except the symbolic one on the top, has the template vector that can by itself perform the template matching between the input (i.e. given from the STM/working memory module) and template vector representing the sound pattern /iːt/ (i.e. the feature vector obtained after the sensory data processing within the **sensation** module(in Chap. 6)). It is then considered that each kernel unit represents

---

[4]At the lexeme level, although the original view of the three visual modality parts i)-iii) agrees with that of the connectionist model by McClelland and Rumelhart (McClelland and Rumelhart, 1981), the auditory counterpart on the other hand corresponds to the so-called TRACE model (McClelland and Elman, 1986), the formation of the former model is fixed, i.e. the structure is not dynamically reconfigurable unlike the one realised by the SOKM (see Chap. 4), and the model is trained via a gradient type method (and hence requires iterative training schemes), whilst the latter (i.e. TRACE) is a rather predefined one (Christiansen and Chater, 1999), i.e. without any learning mechanism equipped to (re-)configure the network. Then, the later connectionist models such as the so-called "simple recurrent networks (SRNs)" (Elman, 1990) (for a general issue of recurrent neural networks, see Mandic and Chambers, 2001) still resort to gradient type algorithms or conventional MLP-NNs (for a survey of the recent models, see Christiansen and Chater, 1999), unlike the models given here.

Related to this, the auditory part of the lexicon has been commonly realised in terms of the hidden Markov models (HMMs) (for a concise review of HMMs for speech applications, see e.g. Rabiner and Juang, 1993; Juang and Furui, 2000). Although it has been reported in many studies that the language processing mechanism modelled by HMMs, e.g. the application to speech recognition, can achieve high recognition accuracy, both the training and testing mostly resort to rather computationally and mathematically complex search (i.e. optimisation) algorithms such as the so-called Viterbi algorithm (Viterbi, 1967; Forney, 1973). Moreover, such higher recognition rates can also be achieved by PNNs (Low and Togneri, 1998). Nevertheless, by means of HMM models, to construct a dynamically reconfigurable system or extend them to multi-modal data processing as realised by the SOKM (in Sect. 4.5) is considered to be very hard.

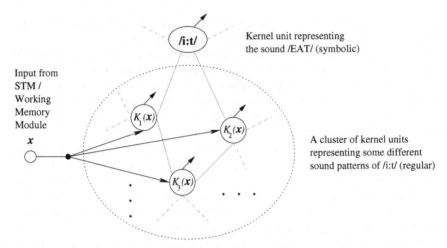

Fig. 9.2. An example of representing the cluster of kernel units for the mental lexicon model – multiple regular kernel units and a symbolic kernel unit representing (or generalising) the sound pattern /i:t/ (/EAT/); it is considered that each kernel unit in the cluster has the template vector that can perform the template matching between the input (i.e. given from the STM/working memory module) and template vector of the sound pattern /i:t/ (Note that, without loss of generality, no specific directional flows between the kernel units are considered in this figure)

and thus generalises to a certain extent a particular sort of sound pattern. In other words, several utterances of a specific speaker could be generalised by a single kernel unit.

In practice, the utility of the symbolic kernel units e.g. the one representing (or generalising) the sound pattern /i:t/ (as depicted on the top in Fig. 9.2) may be dependent upon the manner of implementation; for some applications, it may be convenient to analyse/investigate (by humans) how the data processing within the lexical network actually occurs via the activations by observing the activation states of such symbolic kernel units. (However, in such implementation, it may not be always necessary to introduce actually such symbolic kernel units. In this respect, the same scenario applies to the symbolic kernel units at the conceptual (lemma) level; the concept formation can be simply ascribed to the associations (or the link weights) between the kernel units at the lexeme level.)

Alternatively, it is also considered that the kernel unit on the top of the cluster can be used as the output (or *gating*) node to generalise the activations from the regular kernel units within the cluster, with the activation function, e.g. the linear output given by (3.14), as in the output nodes of PNNs/GRNNs, or the nonlinear one such as the sigmoidal output in (3.29), depending upon the application. Eventually, the transfer of activations can be sent to other domains (or clusters) via such a gating node.

Next, we consider how the data processing within the lexical network as shown in Fig. 9.1 can be actually performed: suppose a situation where a modality-specific data vector, for instance, i.e. the data representing a sound pattern of the word /EAT/, is transferred from the **STM/working memory** module (i.e. due to the receipt of the auditory sensory data after the feature extraction process within the AMS).

Then, as in Fig. 9.1, some of the kernel units within the cluster representing (or generalising) the respective sound patterns (i.e. several different utterances) of the word /i:t/ (/EAT/) can be firstly activated, as well as some of the kernel units within the other clusters, i.e. the clusters of the kernel units representing the respective phonemes /i/, /t/, etc, i.e. depending upon the values of the link weights in between, at the lexeme level.

Second, since some of the kernel units at the lexeme level may have also already established the link weights across different modalities (i.e. due to the data-fusion of the auditory part and that corresponding to the visual modality, occurred during the learning process between the STM/working memory and LTM-oriented modules, as described in Chaps. 7 and 8), the subsequent (or simultaneous) activations from the kernel units in different modalities (i.e. auditory → visual) can also occur (in Chap. 4, we have already seen how such activations can occur via the simulation example of the simultaneous dual-domain (i.e. both the auditory and visual domains) pattern classification tasks by the SOKM).

Then, in the sense that such subsequent activations can occur without actually giving the input of the corresponding modality but due only to the transfer of the activations from the kernel units in other modalities, this simulates the data processing of *mental imagery*.

9.2.2 Concept Formation

Third, this data-fusion can lead to the concept formation at the conceptual (lemma) level, as shown in Fig. 9.1; the emergence of the concept "EAT" can be represented by the activation from the symbolic kernel "EAT" at the lemma level, as well as the subsequent activations from the associated kernel units at both the lemma and lexeme levels, due to the transfer of the activation from the two (symbolic) kernels (or, alternatively, the activations from some of the kernel units at the lexeme level).

For representing the kernel unit "EAT" at the lemma level, it is also considered that, instead of the symbolic kernel unit, a regular kernel unit can be employed, with the input vector $\mathbf{x}_{\text{"EAT"}}$ given as

$$\mathbf{x}_{\text{"EAT"}} = [K_{\text{'EAT'}} \ K_{/i:t/}]^T \tag{9.1}$$

where $K_{\text{'EAT'}}$ (note that here the symbol(s) (i.e. the word(s)) with the expression '·' denotes the image pattern, whereas that in "·" represents the concept) and $K_{/i:t/}$ denote the activation from the kernel unit representing the visual and auditory part of the word "EAT", respectively.

Subsequently, the transfer of the activation from the kernel unit "EAT" can cause other concept formation at the lemma level, e.g. "EAT" → "VERB" and/or "NOUN" ... (needless to say, this also depends upon the strength of the connection, i.e. the current values of the link weights in between), which can eventually lead to the representation of a sentence to be described next. However, to what extent such transfer of the activation is continued depends upon not only the data processing amongst other modules within the AMS but also the current condition of the link weights; in Fig. 9.1, imagine a situation where the kernel unit representing "EAT" at the lemma level is firstly activated (i.e. by the transfer from the lower level kernel unit representing the image pattern 'EAT' $K_{\text{'EAT'}}$, say, due to the input data $\mathbf{x}$ given), then, using (4.3), the activation from the kernel unit representing the concept "HAVE" $K_{\text{"HAVE"}}$ can be expressed by the transfer of the subsequent activations:

$$K_{\text{"HAVE"}} = \gamma w_{\{\text{"HAVE"},\text{"VERB"}\}} \times K_{\text{"VERB"}} \times$$
$$\gamma w_{\{\text{"VERB"},\text{"EAT"}\}} \times K_{\text{"EAT"}} \times$$
$$\gamma w_{\{\text{"EAT"},\text{'EAT'}\}} \times K_{\text{'EAT'}}(\mathbf{x}) . \qquad (9.2)$$

Thus, depending upon the current values of link weights w_{ij}, a certain situation in that the above does not satisfy the relation $K_{\text{"HAVE"}} \geq \theta_K$ (as defined in (3.12)) can be considered, since the subsequent activations from one to another kernel unit are *decaying* due to the factor γ (see Sect. 4.2.2).

9.2.3 Syntax Representation in Terms of Kernel Memory

For describing the concept formation in the previous subsection, it sometimes seems to be rather convenient and sufficient that we only consider the upper level, i.e. the conceptual (lemma) level, without loss of generality; as illustrated in Fig. 9.1, the kernel units at the lemma level can be mostly represented by symbolic nodes rather than regular kernel units. This account also holds for the description of syntax representation. Thus, to describe the syntax representation, or, more generally, language data processing, conventional symbolic approaches are considered to be useful. However, it is seen that, in order to embody such symbolic representation related to the language data processing and eventually incorporate into the design of the AMS, the kernel memory principle can still play the central role. (For instance, various lexical networks based upon conventional symbolism as found in (Kinoshita, 1996) can also be interpreted within the kernel memory principle.)

Then, we here consider how the syntax representation can be achieved in terms of the kernel memory principle described so far. Although to give a full account of the syntax representation is beyond the scope of this book, in this subsection, we see how the principle of kernel memory can be incorporated for the syntax representation.

Now, let us examine a simple sentence, "The dog runs.", by means of the kernel memory representation of the mental lexicon as illustrated in Fig. 9.1:

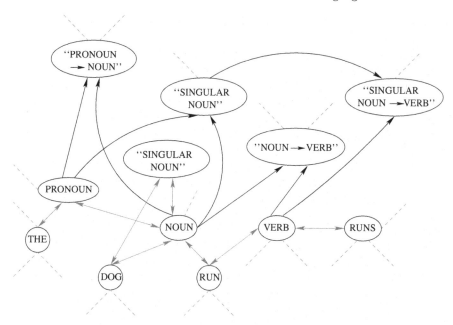

Fig. 9.3. An example of the mental lexicon representing the simple sentence "THE DOG RUNS." in terms of the kernel memory representation.

as in Fig. 9.1, it is firstly considered that the three kernel units representing the respective concepts "THE", "DOG", and "RUN" all reside at the lemma level and can be subsequently activated by the transfer of activations from the kernel unit(s) at the lower (i.e the lexeme) level.

Second, the word order "DOG" then "RUN" can be determined, due to the kernel unit representing the directional flow "NOUN" → "VERB", given the activations from both the (symbolic) kernel units for "DOG" and "RUN" as the input elements (i.e. defined in (9.1)) to the kernel unit, as illustrated in Fig. 9.3 (i.e. since the kernel units for "DOG" and "RUN" have the connection via the link weight with "NOUN" and "VERB", respectively). Similarly, the word order "THE" then "DOG" can be established due to the kernel unit representing the directional flow (or the association in between) "PRONOUN" → "NOUN". (For actually modelling the kernel units that represent such (mono-)directional flows, refer back to Sect. 3.3.4.) Here, it is assumed that these two directional flows, i.e. the flows "NOUN" → "VERB" and "PRONOUN" → "NOUN", have already been acquired through the learning process of the **language** module within the AMS.

Then, it may be seen that the determination of the word order in the above is due to the higher-level concepts such as those represented by the data flow "NOUN" → "VERB" or "PRONOUN" → "NOUN". In other words, the word sequence "THE" → "DOG" → "RUN" follows due to the higher-level concepts formed (in advance) within the lexicon by means of the language module.

However, in contrast to the aforementioned manner of determination, within the context of the learning by the AMS, it is also possible to consider that this has been learnt from the examples; i.e. firstly the concept formation of the words "DOG", "RUNS", "THE", etc, as well as the word sequence, occurs through multiple presentations of such word sequences to the AMS and the associated learning process of the memory modules (see Chaps. 7 and 8). Then, the higher-level concept (i.e. to "generalise" the word sequence) is formed later by a further learning process (e.g. with reinforcement).

Third, similar to the rule of the aforementioned directional flows, it is considered that the rule in which *"since the noun "DOG" is a singular noun of the third person, the following verb must have "S" to indicate this in the present simple form and thus "RUNS", instead of the original "RUN", in English"* has also been acquired through the learning process of the language module (i.e., similar to the higher-level concept of the word sequence, it can be ultimately considered that even this complex rule has been acquired in the aforementioned "learning through examples" principle). This is represented by the sequences of the activations:

1) "THE" → "PRONOUN", "DOG" → "NOUN", and "RUN" → "VERB";
2) The flows in 1) → "SINGULAR NOUN";
3) The flows in 1,2) → "VERB" → "SINGULAR NOUN → VERB" → "RUNS"

Therefore, it can be considered that, within the kernel memory principle, the language module is composed of a set of the grammatical rules which generalises a *chain* of concepts (i.e. represented by a chain of the kernel units responsible for the corresponding concepts, due to the link weights in between *with* directional flows), e.g. "NOUN" → "VERB", "DOG" → "SINGULAR NOUN" → ... "RUNS" ..., and so forth.

Moreover, in (Ullman, 2001; Sakai, 2002), it is considered that the acquisition of the grammatical rules involves the procedural memory, whereas the learning of words is due to the declarative (explicit) LTM. We will return to a further issue of the grammatical rules in terms of the data processing due to the **thinking** module in the next section.

In addition, the utility of the pronoun such as "THE" requires the notion not merely related to the syntactical rules but also (some sort of) the *spatial* information about the AMS (and hence the memory to store it temporarily), i.e. to describe the dog actually exists e.g. in front of the body (thus, the dog is "spatially" away and perceived via the input: sensation module), or to remember (shortly) the concept of the "dog" that specifies a certain dog appeared previously in the context (thus, the requirement for the temporal memory). Therefore, it is considered that the notion of the pronouns such as

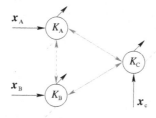

Fig. 9.4. A kernel (sub-)network consisting of the three kernel units K_A, K_B, and K_C

"IT", "THAT","THIS", etc, also involves the data processing within other modules (such as the **memory** /innate structure modules) of the AMS.

Before moving on to the discussion of the thinking module, we revisit the issue of how the concept formation can be realised within the kernel memory context in the next subsection, which is also closely related to the implementation of the syntax representation described so far.

9.2.4 Formation of the Kernel Units Representing a Concept

In Sect. 3.3.4, it was described how a kernel unit can represent the directional flow between a pair of kernel units. In a similar context, we here consider how the kernel units representing a concept can be formed within the SOKM principle (in Chap. 4).

Now, let us consider the following scenario:

i) A kernel unit K_A is added into the memory space, at time index $n = n_1$ (i.e. by following the [**Summary of Constructing A Self-Organising Kernel Memory**] on page 63)
ii) Another kernel unit K_B is then added, at $n = n_2$;
iii) Next, the kernel unit K_C representing a certain concept that can be related to the two added kernel units K_A and K_B is added, at $n = n_3$;
iv) The links between the kernel units K_C and K_A/K_B are formed at $n = n_4$ (i.e. $n_1 < n_2 < n_3 < n_4$).

Thus, at time $n = n_4$, it is considered that the kernel (sub-)network as shown in Fig. 9.4 is formed. (In the figure, note that the respective inputs to the three kernel units $\mathbf{x}_A$, $\mathbf{x}_B$, and $\mathbf{x}_C$ are not necessarily those belonging to the same domain.) In Fig. 9.4, it is possible to consider such a situation that (during the early stage of the memory construction) the link weight between K_A and K_B is formed at $n > n_2$. Then, the (sub-)network structure in Fig. 9.4 is *tournament*, in the sense that each node is connected by bi-directional links between all the three kernels and can be simultaneously activated due to the transfer of the activation from any kernel(s).

The Kernel Unit Representing a Directional Flow

Now, consider a situation where the kernel unit K_C represents a certain concept that can be activated by the sequential activation of K_A and K_B, i.e. representing the directional flow $K_A \rightarrow K_B$ as in Sect. 3.3.4. In such a situation, it is considered that, although initially the link weight between K_A and K_B was formed, the link weight w_{AB} may eventually disappear, i.e. according to [**The Link Weight Update Algorithm**] (on page 60) followed by Conjecture 1 (on page 60) within the SOKM principle; unless the simultaneous excitation of K_A and K_B occurs periodically, the value of the link weight in between will be decreased in time (i.e. denoted by the *grey-coloured link* in between in Fig. 9.4). Instead, by extending Conjecture 1 within the SOKM context and exploiting the template matrix for the temporal representation as in (3.32), we may draw the following conjecture:

Conjecture 4: When a pair of kernels K_i and K_j $(i \neq j)$ in the SOKM are *asynchronously* and repeatedly excited, a new kernel unit K_{new} representing the asynchronous excitation between K_i and K_j may be formed, where appropriate, with its input

$$\mathbf{X}_{new}(n) = \begin{bmatrix} K_i(n) \ K_i(n-1) \ \dots \ K_i(n-p_{new}+1) \\ K_j(n) \ K_j(n-1) \ \dots \ K_j(n-p_{new}+1) \end{bmatrix}, \quad (9.3)$$

where $K_{i/j}(n)$ denotes the activation of the kernel unit $K_{i/j}$ at time n, and the template matrix:

$$\mathbf{T}_{new} = \begin{bmatrix} t_i(1) \ t_i(2) \ \dots \ t_i(p_{new}) \\ t_j(1) \ t_j(2) \ \dots \ t_j(p_{new}) \end{bmatrix}. \quad (9.4)$$

where the element $t_{i/j}(k)$ $(k = 1, 2, \dots, p_{new})$ may be alternatively given by (3.33) or (3.34) and p_{new} is a positive constant.

Note that Conjecture 4 may be seen as an alternative representation of the directional flow between a pair of kernel units that is useful to know the exact timing of occurring such a directional data flow in between, where required for further data processing (thus, to justify the biological plausibility is beyond the scope of this book).

Then, we exploit Conjecture 4 in the above for the formation of a new kernel unit K_{AB}, as shown in Fig. 9.5; in the figure, the new kernel unit K_{AB} is formed (at $n > n_4$) with the input

$$\mathbf{X}_{AB}(n) = \begin{bmatrix} K_A(n) \ K_A(n-1) \ \dots \ K_B(n-p_{AB}+1) \\ K_B(n) \ K_B(n-1) \ \dots \ K_B(n-p_{AB}+1) \end{bmatrix}. \quad (9.5)$$

Note that, at this point, since the connections between the kernel units K_{AB} and K_A/K_B are represented in terms of the input vector to the kernel

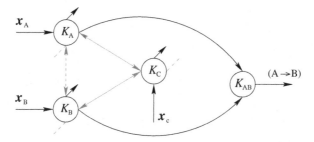

Fig. 9.5. Formation of a new kernel unit K_{AB} which represents the directional flow between the two kernel units $K_A \rightarrow K_B$ within the (sub-)network (formed at $n = n_4$)

K_{AB}, rather than the ordinary (bi-directional) link weights, the data flow in reverse, i.e. $K_{AB} \rightarrow K_A, K_B$, is not allowed.

Accordingly, the template matrix for the kernel unit K_{AB} is represented as in (9.4):

$$\mathbf{T}_{AB} = \begin{bmatrix} t_A(1) \ t_A(2) \ \dots \ t_A(p_{AB}) \\ t_B(1) \ t_B(2) \ \dots \ t_B(p_{AB}) \end{bmatrix}. \tag{9.6}$$

It is then considered that, as described in Sect. 3.3.4, the kernel unit K_{AB} can eventually represent the directional flow of $K_A \rightarrow K_B$ by varying (due to the associated learning process) either the number of columns p_{AB} (see page 55) or the regularisation factor κ_A/κ_B (see page 56).

Establishment of the Link Weight Between K_{AB} and K_C – the Concept Formation

As described in the previous subsection, after a certain learning process of the kernel K_{AB} to represent the directional flow $K_A \rightarrow K_B$, K_{AB} can be activated if the pattern matching between an asynchronous activation pattern of K_A and K_B and the template matrix $\mathbf{T}_{AB}$ is successfully done (see Sect. 3.3.4).

In such a situation, it is considered that both the kernels K_{AB} and K_C can be subsequently (or simultaneously) activated, if the link between these two kernels is already established (i.e. during the associated learning process).

Figure 9.6 shows the case where the bi-directional link between K_{AB} and K_C is established within the sub-network shown in Fig. 9.5. Then, the following two cases of the activation for K_{AB} and K_C are considered:

1) The kernel unit K_{AB} is firstly activated due to the asynchronous activation between K_A and K_B (i.e. given as the input to K_{AB}), and then the activation from K_{AB} is transferred, which causes the subsequent activation from the kernel K_C.

2) In reverse, the kernel unit K_C is firstly activated by its input $\mathbf{x}_c$ or the transfer via the link weight(s) from the kernel unit(s) other

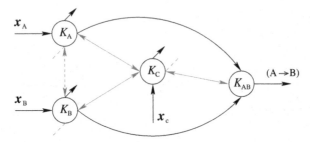

Fig. 9.6. Establishment of the bi-directional link between K_{AB} and K_C within the (sub-)network

than those within the sub-network, and then the activation from
K_{AB} subsequently occurs.

In both the cases above, it is also possible that both the kernels K_A and
K_B can be eventually activated due to the subsequent transfer of the activation from K_C, which may be exploited further for simulating the imagery
task, e.g. to recover the constituents of the concept.

Nevertheless, it is macroscopically viewed that the transfer of activation
occurring at the sub-network is related to the concept formation that represents the directional flow $K_A \rightarrow K_B$ within the SOKM context.

Extension to the Case for More Than Two Constituent Kernel Units Involved

In the previous case as shown in Fig. 9.6, only two kernel units were involved
for the concept formation. Here, we consider how this principle can be generalised to the case where more than two constituent kernel units are involved.

Figures 9.7 and 9.8 show the two possible situations where the kernel units
$K_1, K_2, \ldots, K_N$ are the constituents to form the concept that is represented
by the kernel K_C^N. In Fig. 9.7, it is considered that the kernel unit K_C^N represents the flow of $K_1 \rightarrow K_2 \rightarrow \ldots \rightarrow K_N$ *at a time*; the template matrix has
the size $(N \times p_{C^N})$. In contrast, the network structure in Fig. 9.8 shows the
case where each kernel unit K_C^i $(i = 2, 3, \ldots, N)$ represents the subsequent
directional flow, i.e. K_C^2: $K_1 \rightarrow K_2$, K_C^3: $K_1 \rightarrow K_2 \rightarrow K_3$, and, eventually,
the kernel unit K_C^N represents the directional flow of $K_1 \rightarrow K_2 \rightarrow \ldots \rightarrow K_N$,
all with the template matrix of size $(2 \times p_{C^i})$.

In this section, so far a framework of how the language and semantic net-
works/lexicon modules can be represented has been given based upon the kernel memory principle. However, the description has been rather restricted to
the structural sense of these language-oriented modules. In the following section, we consider the thinking module, which incorporates with the other associated modules within the AMS, and see how the interactive data processing

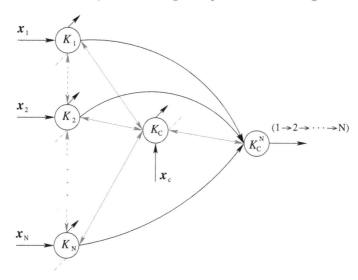

Fig. 9.7. Concept formation involving more than two constituent kernel units – the kernel unit K_C^N represents the directional flow of $K_1 \rightarrow K_2 \rightarrow \ldots \rightarrow K_N$, at a time

amongst such modules also involves and then contributes to reconfigure the language-oriented modules.

9.3 The Principle of Thinking – Preparation for Making Actions

In Sect. 9.2.3, it has been described how each lexeme can be organised to form eventually a sentence in terms of the kernel memory principle by examining through the example of the simple sentence "The dog runs.". Then, consider another example similar to the previous, i.e. "Dog flies." According to the principle of the word sequence described earlier, this sentence can be found to be grammatically correct. However, the correctness in the sense of "meaning" (or the *semantic* sense) of this sentence depends upon the context/situation in which the sentence is used. Therefore, it is considered that one of the roles for the **thinking** module is to judge the correctness in the semantical sense.

As shown in Fig. 5.1, the thinking module within the AMS is considered to function in parallel with the **STM/working memory** module (i.e. the connection between the two modules is depicted (*solid line*) without arrows in Fig. 5.1) and as a mechanism to organise the data processing with the four associated modules, i.e. 1) **intention**, 2) **intuition**, and 3) **semantic networks/lexicon** module.

Then, provided that the subsequent activations from the kernel units corresponding to the sentence, "Dog flies.", occur within the semantic networks/lexicon module, e.g. given appropriate stimuli(us) to the AMS, it is

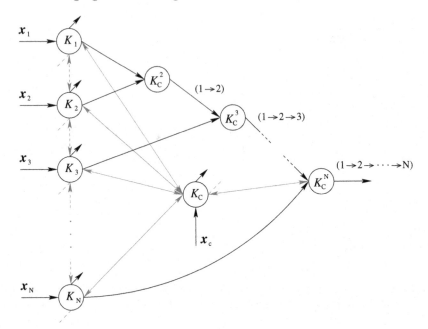

Fig. 9.8. Concept formation involving more than two constituent kernel units – each kernel unit K_C^i ($i = 2, 3, \ldots, N$) represents the directional flow subsequently; i.e. K_C^2: $K_1 \rightarrow K_2$, K_C^3: $K_1 \rightarrow K_2 \rightarrow K_3$, and, eventually, the kernel unit K_C^N represents the directional flow of $K_1 \rightarrow K_2 \rightarrow \ldots \rightarrow K_N$

considered that the interactive data processing amongst these four modules and the STM/working memory module occurs, in order to determine whether the sentence is semantically correct or not.

More concretely, provided that the AMS has already acquired the fact that "Dog runs but does not fly in reality" (i.e. in terms of the explicit LTM) during the learning process; i.e. during the exposition of the AMS to the surrounding environment, the AMS must have not encountered any situation in which the dog flies in reality, or due to the reinforcement given (e.g. the teacher's signal) during its learning process (in Chap. 7). Then, it can be considered that the corresponding association process (i.e. the data-fusion) between the concepts "DOG" and "FLY" did not occur.

However, it still is possible to consider such a situation where the sentence, "Dog flies.", can appear in virtual reality e.g. in a fantasy novel, and the AMS acquired such knowledge (in terms of the kernel memory representation) through the associated learning process. In such a situation, it is considered that the thinking module within the AMS performs the memory search to a certain extent and eventually, for instance, contributes to accomplish the following sequence: "VIRTUAL WORLD" → "DOG" → "FLIES", by accessing the (episodic) contents of the LTM.

Thus, it is said that the principal role of the thinking module is to perform the memory search multiple times (i.e. within the kernel memory so constructed) and that the manner of such search processes is quite dependent upon the current states of the other modules associated with both the thinking and STM/working memory modules (i.e. since both the two modules function in parallel) such as **emotion**, intention, and/or intuition module.

9.3.1 An Example of Semantic Analysis Performed via the Thinking Module

Now, to be more concrete, let us have a closer look at the simple example of lexical analysis of the sentence, "Dog flies.", performed via the thinking module: as described earlier, suppose first that the AMS successfully processes the sensory data (i.e. the utterance spoken by a human through the microphone(s)) and then that the sequence of the two words is found to be grammatically correct by the AMS, i.e. by accessing the kernel units within the semantic networks/lexicon module (for the detail, refer back to Sects. 9.2.1 and 9.2.2).

Second, the semantic analysis can be performed via the thinking module; e.g. by the **STM/working memory** module, the activated kernel units corresponding to the word sequence (i.e. the kernel units both/either at the lemma and/or at the lexeme level) are marked, and the states of such activations are preserved within the STM/working memory or the **intention** module for a certain (short) period of time. (At this point, it is also possible to consider that there are other kernel units within the STM/working memory that are irrelevant to the word sequence, e.g. other incoming sensory data.) Then, with the current states in both the **attention** and **emotion** modules, the semantic analysis starts by accessing (several times) the episodic contents of the **LTM** modules (e.g. searching the kernel unit(s) storing the image(s) of "dog flies") or the **intuition** module (i.e. the latter module will be described later). (Note that this analysis can also involve the memory search within the **semantic networks/lexicon** module, due to the link weight connections in between.)

Third, during the search process within the episodic memory, if the subsequent activations from the kernel units representing the three concepts "DOG", "FLIES", and "VIRTUAL WORLD" eventually occur (repetitively or consistently), such kernel units together with the link weight connections formed temporarily can remain within the STM/working memory module. Then, if the activations from such kernel units last and are marked by the STM/working memory module for a sufficiently long period of time, the kernel network representing the subsequent concept formation may eventually turn to the network of the corresponding LTM module(s) (even though it is also possible to consider that such formation of the kernel network representing the subsequent concepts may be later altered or disappear completely from the LTM modules due to the further learning process, i.e. by the reinforcement).

Moreover, it is possible that, a new/further search is triggered from such kernel units to represent other concepts and starts a new thinking process via the thinking module.

In the next chapter, the four modules associated with the abstract notions 1) attention, 2) emotion, 3) intention, and 4) intuition module, all of which have appeared in the example above, will be described in detail.

9.3.2 The Notion of Nonverbal Thinking

In the previous subsection, one of the fundamental roles of the thinking module, namely the semantic analysis of the sequence of words via the language module and the associated LTM structures, has been described. However, it is generally considered (albeit dependent upon the manner of interpretation) that the thinking process is performed not only verbally but also nonverbally (cf. e.g. Sakai, 2002).

In the AMS context, it can be seen that the verbal thinking process corresponds to such a process as in the previous example of the semantic analysis, whereas the nonverbal thinking is the process in which the kernel units not at the lemma level but at the lexeme level significantly contribute to the memory search by the thinking module. In this sense, such memory search is also related to describe the notion of **intuition** to be discussed in the next chapter.

9.3.3 Making Actions – As a Cause of the Thinking Process

Regardless of verbal or nonverbal thinking processes, it is considered that the activations from some of the kernel units within the LTM/LTM-oriented modules can induce the subsequent activations (i.e. due to the link weight connections) from those which are directly connected to control the respective mechanisms in the body via the **primary outputs** module. (In Fig. 5.1, this is indicated by the mono-directional flow from the **implicit LTM** to the primary output module.)

It is also considered that, as shown in Fig. 5.1 (on page 84), the internal states due to the procedural memory part of the implicit LTM module can be varied, during the memory search process via the thinking module.

9.4 Chapter Summary

In this chapter, it has been described that both the language and thinking modules are closely tied to each other within the AMS context. Then, a framework for designing both the language and thinking modules has been given, by examining through several examples based upon the kernel memory principle. As described earlier, although to obtain a complete picture of these two modules still requires a further study in relevant disciplines, such as linguistics or developmental psychology, as well as more design-oriented studies e.g. in AI or robotics, the kernel memory representations have been demonstrated still to play the central role in the actual design of the two modules.

As described, it can be seen that the language module consists of a set of grammatical rules and incorporates with the thinking module to form the sentences, whilst the thinking module functions in parallel with the **STM/working memory** and plays the role in the interactive data processing amongst the three associated modules, i.e. 1) **intention**, 2) **intuition**, and 3) **semantic networks/lexicon** module, with/without the language-oriented data processing (i.e. corresponding to the verbal/nonverbal thinking). It is considered that the thinking process (i.e. regardless of the verbal or nonverbal processes) may eventually invoke real actions by the body via the **primary output** module. As shown in Fig. 5.1, this can happen due to the accesses and thereby the subsequent activations within the **implicit LTM** module, during the memory search process, via the thinking module.

In the next chapter, we move on to the discussion of the remaining four modules associated with the abstract notions related to the mind, namely, the attention, emotion, intention, and intuition modules.

10

Modelling Abstract Notions Relevant to the Mind and the Associated Modules

10.1 Perspective

This chapter is devoted to the remaining four modules within the AMS, i.e. 1) **attention**, 2) **emotion**, 3) **intention**, and 4) **intuition** module, and their mutual interactions with the other associated modules. Then, the four modules so modelled represent the respective abstract notions related to the mind.

10.2 Modelling Attention

In the late nineteenth century, the psychologist William James wrote (James, 1890):

> "Everyone knows what attention is. It is the taking possession by the mind, in clear and vivid form, of one out of what seem several simultaneously possible objects or trains of thought. Focalization, concentration, of consciousness are of its essence. It implies withdrawal from some things in order to deal effectively with others, and is a condition which has a real opposite in the confused, dazed, scatterbrain state...."

and his general notion of "attention", after more than one hundred and fifteen years, is still convincing in various modern studies relevant to general brain science such as cognitive neuroscience/psychology (Gazzaniga et al., 2002).

In psychology, despite proposals of a variety of (conceptual) connectionist models for selective attention, such as the "selective attention model" (SLAM) (Phaf et al., 1990), "multiple object recognition and attentional selection" (MORSEL) (Mozer, 1991; Mozer and Sitton, 1998) or "selective attention for identification model" (SAIM) (Heinke and Humphreys, in-press), and for a survey of such connectionist models (see Heinke and Humphreys, in-press), little has been reported for the development of concrete models of attention and their practical aspects.

Tetsuya Hoya: *Artificial Mind System – Kernel Memory Approach*, Studies in Computational Intelligence (SCI) **1**, 189–235 (2005)
www.springerlink.com

In the study (Gazzaniga et al., 2002), the function of "attention" is defined as "*a cognitive brain mechanism that enables one to process relevant inputs, thoughts, or actions, whilst ignoring irrelevant or distracting ones*".

Then, within the AMS context, the notion of attention generally agrees with that in the aforementioned studies; as indicated in Fig. 5.1 (i.e. by the bi-directional data flows, on page 84), it is considered that the **attention** module primarily operates on the data processing within both the **STM/working memory** and **intention** modules. The attention module is also somewhat related to the **input: sensation** module (i.e. this is indicated by the link between the attention and input: sensation module shown (*dashed line*) in Fig. 5.1), since, from another point of view, some pre-processing mechanisms within the sensation module such as BSE, BSS, DOA, NR, or SAD, can also be regarded as the respective functionalities dealt within the notion of attention; for instance, the signal separation part of the blind speech extraction models, which simulates the human auditory attentional system in the so-called "cocktail party situations" (as described extensively in Sect. 8.5), can be treated as a pre-processing mechanism within the sensation module. (In this sense, the notion of the attention module within the AMS also agrees with the cognitive/psychological view of the so-called "early-versus late-selection" due to the study by Broadbent (Broadbent, 1970; Gazzaniga et al., 2002).)

10.2.1 The Mutual Data Processing:
Attention ⟷ STM/Working Memory Module

For the data processing represented by the data flow **attention ⟶ STM/ working memory** module, it is considered that the attention module functions as a *filter* which picks out a particular set of data and then holds temporarily its information such as i.e. the activation pattern of some of the kernel units within the memory space, e.g. due to a subset of the sensory data arriving from the **input: sensation** module, amongst the flood of the incoming data, whilst the rest are bypassed (and transferred to e.g. the **implicit LTM** module; in due course, it can then yield the corresponding perceptual outputs), the principle of which agrees with that supported in general cognitive science/psychology (see e.g. Gazzaniga et al., 2002), so that the AMS can efficiently and intensively perform a further processing based upon the data set so acquired, i.e. the thinking process.

Thus, in terms of the kernel memory context, the attention module urges the AMS to set the current *focus* to some of the kernel units, which fall in a particular domain(s), amongst those within the STM/working memory module as illustrated in Fig. 10.1, (or, in other words, the priority is given to *some* (i.e. not all) of the marked kernel units in the entire memory space by the STM/working memory module; see Sect. 8.2), so that a further memory search process can be initiated from such "attended" kernel units, e.g. by the associated modules such as **thinking** or **intention** modules, until the current focus is switched to another. (In such a situation, the attention module

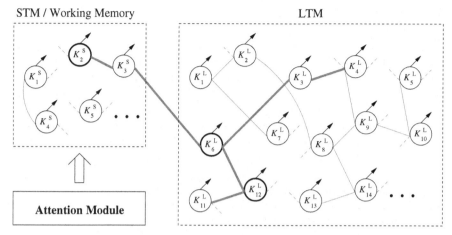

STM / Working Memory

LTM

Fig. 10.1. An illustration of the functionality relevant to the attention module –
focusing upon some of the kernel units (i.e. the "attended" kernel units) within
the **STM/working memory** and/or **LTM** modules, in order to urge the AMS to
perform a further data processing relevant to a particular domain(s) selected via the
attention module, e.g. by the associated modules such as **thinking** or **intention**
module (see also Fig. 5.1); in the figure, it is assumed that the three activated
kernel units K_2^S, K_6^L, and K_{12}^L (*bold circles*) within the STM/working memory (i.e.
the former kernel unit) and LTM modules (i.e. the latter two) are firstly chosen as
the *attended* kernel units by the attention module. Then, via the link weights (*bold
lines*), the activations from some of the connected kernel units can subsequently
occur within the LTM modules (Note that, without loss of generality, no specific
directional flows between the kernel units are considered in this figure)

temporarily holds the information about e.g. the locations of the kernel units
so marked.)

More concretely, imagine a situation that now the current focus is set to
the data corresponding to the voiced sound uttered by a specific person and
then that some of the kernel units within the associated memory modules are
activated by the transfer of the incoming data corresponding to the utterances
of the specific person and marked as the *attended* kernel units. (In Fig. 10.1,
the three kernel units K_2^S, K_6^L, and K_{12}^L correspond to such attended kernel
units.) Then, although there can be other activated kernel units which are
marked by the STM/working memory module but irrelevant to the utterances, a further data processing can be invoked by the thinking module with
priority; e.g. prior to any other data processing, the data processing related to
the utterances by the specific person, i.e. the grammatical/semantic analysis
via the **semantic networks/lexicon**, **language**, and/or thinking module, is
mainly performed, due to the presence of such attended kernel units (i.e. this
is illustrated by the link weight connections (*bold lines*) in Fig. 10.1). Moreover, it is also possible to consider that the **perception** of other data (i.e.

due to the PRS within the implicit LTM) may be intermittently performed in parallel with the data processing.

In contrast to the effect of the attention module upon the STM/working memory module, the inverted data flow **STM/working memory** ⟶ **attention** module indicates that the focus can also be varied due to the indirect effect from the other associated modules such as the **emotion** or thinking modules, via the STM/working memory module. More specifically, it is possible to consider a situation where, during the memory search process performed by the thinking module, or due to the flood of sensory data that fall in a particular domain(s) arriving at the STM/working memory module/the memory recall from the LTM modules, the activated kernel units representing the other domain(s) may become more dominant than that (those) of the initially attended kernel units. Then, the current focus can be greatly affected and eventually switched to another.

Similarly, the current focus can be greatly varied due to the emotion module via the STM/working memory module, since the range of the memory search can also be significantly affected, due to the current emotion states within the emotion module (to be described in the next section) or the other internal states of the body.

10.2.2 A Consideration into the Construction of the Mental Lexicon with the Attention Module

Now, let us consider how the concept of the attention module is exploited for the construction of the mental lexicon as in Fig. 9.1 (on page 172)[1].

As in the figure, the mental lexicon consists of multiple clusters of kernel units, each cluster of which represents the corresponding data/lexical domain and, in practice, may be composed by the SOKM principle (i.e. described in Chap. 4).

Then, imagine a situation where, at the lexeme level, the clusters of the kernel units representing elementary visual feature patterns or phonemes are firstly formed within the implicit LTM module (or, already pre-determined, in respect of the innateness/PRS, though they can be dynamically reconfigured later during the learning process), but where, at the moment, those for higher level representations, e.g. the kernel units representing words/concepts, still are not formed.

Second, as described in Chap. 4, the kernel units for a certain representation at the higher level (i.e. a cluster of the kernel units representing a word/concept) are about to be formed from scratch within the corresponding LTM module(s) (i.e. by following the manner of formation in [**Summary of Constructing A Self-Organising Kernel Memory**] on page 63) and

---

[1]Although the model considered here is limited to both the auditory and visual modalities, its generalisation to multi-modal data processing is, as aforementioned, straightforward within the kernel memory context.

eventually constitute several kinds of kernel networks, *due to the focal change by the attention module.*

Then, as described in Sect. 9.2.2, the concept formation can be represented based upon the establishment of the link weight(s) between the newly formed kernel units (at the higher level) and those representing elementary components (at the lower level), via the focal change due to the attention module. (Alternatively, within the kernel memory context, such concept formation can be represented, without defining explicitly such distinct two levels and then establishing the link weights between the two levels, but rather by the data directly transferred from the STM/working memory module; i.e. a single kernel unit is formed and stores [a chunk of] the modality specific data within the template vector, e.g. representing a whole word at a time.)

Related to the focal change, it may also be useful/necessary to take into account the construction of a hierarchical memory system for the efficiency in terms of the computation; as illustrated in Fig. 8.2 (on page 154), the subsequent pattern recognition (i.e. **perception**) processes must be quickly performed, in order to deal with the incessantly varying situation encountered by the AMS (i.e. this is always performed to seek the rewards or avoid the obstacles, resulting from the **innate structure** module). Thus, depending upon the current situation perceived by the AMS, the attention module will change the focus. (For this change, not solely the attention module but also other modules, i.e. the **intention**, **emotion**, and/or **thinking** modules, can therefore be involved.)

In addition to this, from a linguistic point of view, it may be said that the memory hierarchy as in Fig. 8.2 may follow the so-called "difference structure", due to the great French thinker, Ferdinand-Morgin de Saussure (for a comprehensive study/concise review of his concepts, cf. e.g. Maruyama, 1981); e.g. from the sequences of words, "the dog", "the legs", "the person" ..., the concept of the single word representing the definite article "the" can be detached from the word sequences and formed, with the aid of the attention module.

More concretely, provided that the auditory data of the sequences of the words are, for instance, stored in advance within the respective template vectors of kernel units within the LTM, it can be considered that, due to the focal change by the attention module, the kernel units, i.e. each with the template vector of shorter length representing the respective utterances of the single word "the", can later be formed (in terms of the kernel memory principle). Then, it is considered that the link weight connections between the kernel units representing the respective sequences of the words and those representing the single word "the" are eventually established.

10.3 Interpretation of Emotion

In general cognitive science, the notion of emotion is regarded as a psychological state or process in order to vary the course of action and eventually achieve certain goals, elicited by evaluating an event as relevant to a goal (Wilson and Keil, 1999). The study of emotion has its own rich history and even backdates to the philosophical periods of time due to Aristotle and Descartes (e.g. Descartes, 1984-5) to the evolutionary study by Darwin (Darwin, 1872)/the psychological studies James (James, 1884) and Freud (see e.g. Freud, 1966) to a modern cognitive scientific insight initiated by Bowlby in the 1950's (see e.g. Bowlby, 1971) and then built upon by many more recent researchers (e.g. Arnold and Gasson, 1954; Schachter and Singer, 1962; Tomkins, 1995).

Then, it is considered that the notion of emotion can be distinguished in time-wise into 1) *affection*, 2) *mood*, and 3) *personality traits* (Oatley and Jenkins, 1996; Wilson and Keil, 1999); the first (i.e. *affection*) is often associated with brief (i.e. lasting a few seconds) expressions of face and voice and with perturbation of the autonomic nervous system, whilst the latter two last relatively longer, i.e. a *mood* tends to resist (temporarily) disruption, whereas the *personality traits* last for years or a lifetime of the individual.

In psychiatric studies (Papez, 1937; MacLean, 1949, 1952), the limbic system, i.e. consisting of the real brain regions including the hypothalamus, anterior thalamus, cingulate gyrus, hippocampus, amygdala, orbitofrontal cortex, and portions of the basal ganglia, is considered to play a principal role in the emotional processing (for a concise review, see e.g. Gazzaniga et al., 2002), though the validity of their concept has still been under study (Brodal, 1982; Swanson, 1983; Le Doux, 1991; Kotter and Meyer, 1992; Gazzaniga et al., 2002). Nevertheless, in the present cognitive study, the general notion is that emotion is not involved in only a single neural circuit or brain system but rather is a multifaceted behaviour relevant to multiple brain systems (Gazzaniga et al., 2002).

In contrast to the aforementioned issues of the brain regions, there has been another line of studies, i.e. rather than focusing upon specific brain systems relevant to the emotional processing, investigating how the left and right hemispheres of the brain mutually interact and eventually contribute to the emotional experience (Bowers et al., 1993; Gazzaniga et al., 2002). For instance, in the neuropsychological study by Bowers et al. (Bowers et al., 1993; Gazzaniga et al., 2002), it is suggested that the right hemisphere is more significant for communication of emotion than the left hemisphere, the notion of which has been supported by many neuropsychological studies of the patients with brain lesions (e.g. Heilman et al., 1975; Borod et al., 1986; Barrett et al., 1997; Anderson et al., 2000).

10.3.1 Notion of Emotion within the AMS Context

As indicated in Fig. 5.1, the **emotion** module within the AMS functions in parallel with the three modules, i.e. 1) **instinct: innate structure**, 2) **explicit/implicit LTM**, and 3) **primary output** module (in Fig. 5.1, all denoted by the respective links in between, on page 84).

In terms of the relations with the 1) instinct: innate structure and 3) primary output modules, it is implied that the emotion module exhibits the aspect of innateness; the emotion module consists of some state variables which represent (a subset of) the current internal states related to the AMS/body and directly reflect e.g. the electrical current flow within the body (thus the module can also be regarded as one of the primary outputs, simulating the elicitation of autonomic responses, such as a change in the heart rate/endocrines, or releasing the stress hormones in the organism (cf. Rolls, 1999; Gazzaniga et al., 2002)) in order to keep the balance.

On the other hand, the functionality in parallel with the 2) explicit/implicit LTM module implies the memory aspect of the emotion module; some of the kernel units in these LTM modules may also have connections via the link weights with the state variables within the emotion module. Figure 10.2 illustrates the manner of connections between the emotion and memory modules within the AMS.

In the figure, it is assumed that the state variables $E_1, E_2, \ldots, E_{N_e}$ have connections with the three kernel units within the memory modules, i.e. K_5^S within the STM/working memory, K_{11}^L and K_{14}^L within the LTM module, via the link weights in between. In such a case, the state variables $E_1, E_2, \ldots, E_{N_e}$ may be represented by symbolic kernel units (in Sect. 3.2.1).

Then, as described earlier, the weighting values represent the strengths between the (regular) kernel units within the memory modules and state variables, which may directly reflect, e.g. the amount of such current flow to change the internal states of the body (i.e. representing the endocrine) via the primary output module.

Alternatively, the kernel unit representation shown in Fig. 10.3 (i.e. modified from Hoya, 2003d) can be exploited, instead of the ordinary kernel unit representations in Figs. 3.1 (on page 32) and 3.2 (on page 37); the (emotional) state variables attached to each kernel unit can be used to determine the current internal states.

10.3.2 Categorisation of the Emotional States

In our daily life, we use the terms such as *angry, anxious, disappointed, disgusted, elated, excited, fearful, guilty, happy, infatuated, joyful, pleased, sad, shameful, smitten,* and so forth, to describe the emotional experience. However, it is generally difficult to translate these into discrete states. In general cognitive studies, there are two major trends to categorise such emotional expressions into a finite set (for a concise review, see Gazzaniga et al., 2002);

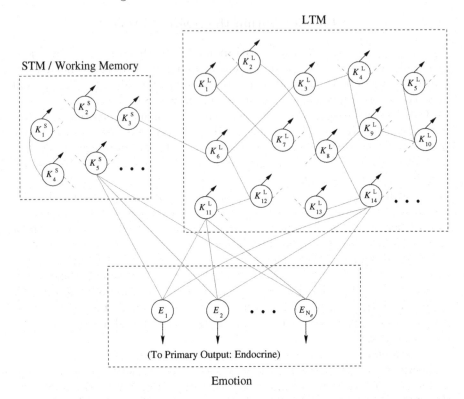

Fig. 10.2. Illustration of the manner of connections between the emotion and memory modules within the kernel memory context by exploiting the link weights in between; in the figure, three kernel units, i.e. K_5^S within the STM/working memory, K_{11}^L and K_{14}^L both within the LTM module, have the connections via the link weights in between with the state variables $E_1, E_2, \ldots, E_{N_e}$ within the emotion module (without loss of generality, no specific directional flows are considered between the kernel units in this figure). Note that such state variables can be even regarded as symbolic kernel units within the kernel memory context. Then, the changes in the state variables directly reflect the current internal states of the body via the **primary output** module (i.e. endocrine)

one way is to characterise basic emotions by examining the universality of the facial expressions of humans (Ekman, 1971), whilst the other is the so-called dimensional approach by describing the emotional states as not discrete but rather reactions to events in the world that vary along a continuum. For the former approach, the four (e.g. amusement, anger, grief, and pleasure) (see e.g. Yamadori, 1998) or six (e.g. those representing anger, fear, disgust, grief, pleasure, and surprise) (cf. Ekman, 1971) emotional states are normally considered, whilst the latter is based upon the two factors, i.e. i) valance (i.e. pleasant-unpleasant or good-bad) and ii) arousal (i.e. how intense is the internal emotional response, high-low) (Osgood et al., 1957; Russel, 1979), or

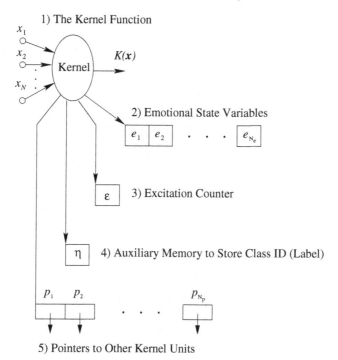

1) The Kernel Function

2) Emotional State Variables

3) Excitation Counter

4) Auxiliary Memory to Store Class ID (Label)

5) Pointers to Other Kernel Units

Fig. 10.3. The modified kernel unit with the emotional state variables $e_1, e_2, \ldots, e_{N_e}$ (i.e. extended from Hoya, 2003d)

(more cognitive sense of) motivation (i.e. approaching-withdrawal) (Davidson et al., 1990).

Similar to the dimensional approaches, in (Rolls, 1999), it is proposed that the emotions should be described and classified according to whether the reinforcer is positive or negative; the emotional states are described in terms of the 2D-diagram, where there are two orthogonal axes representing the respective intensity scales of the emotions associated with the reinforcement contingencies; i.e. the horizontal axis goes in the direction of positive reinforcer ($\overline{S+}$ or S+!) → negative reinforcer ($\overline{S-}$ or S-!), indicating the omission/termination level of the reinforcer (e.g. rage, anger/grief, frustration/sadness, and relief), whilst the vertical axis goes in a similar fashion (i.e. from (S+) to (S-)), showing the presentation level of the reinforcer (e.g. ecstasy, elation, pleasure, apprehension, fear, and terror), and the intersection of these two axes represents the neutral state.

Although so far a number of approaches to define emotions have been proposed, there is no single correct approach (Gazzaniga et al., 2002).

Nevertheless, within the AMS context, it is considered that the emotional states can be sufficiently represented by exploiting the multiple state variables as in Figs. 10.2 and 10.3, depending upon the application, since the objective here is limited to imitating the emotions of creatures and the resultant behaviours.

As an example, we may simply assign the two emotional states E_1 and E_2 in Fig. 10.2 (or the emotional state variables e_1 and e_2 attached to the kernel units in Fig. 10.3) to the respective intensity scales representing the emotions due to Rolls (Rolls, 1999): e.g.

$$E_1(\text{or } e_1) = \begin{cases} a_1 & : \text{ecstasy} \\ a_2 & : \text{elation} \\ a_3 & : \text{pleasure} \\ a_4 & : \text{(neutral)} \\ a_5 & : \text{apprehension} \\ a_6 & : \text{fear} \\ a_7 & : \text{terror} \end{cases} \qquad (10.1)$$

where $a_1 > a_2 > \ldots > a_7$, and

$$E_2(\text{or } e_2) = \begin{cases} b_1 & : \text{rage} \\ b_2 & : \text{anger/grief} \\ b_3 & : \text{frustration/sadness} \\ b_4 & : \text{(neutral)} \\ b_5 & : \text{relief} \end{cases} \qquad (10.2)$$

where $b_1 > b_2 > \ldots > b_5$.

Then, the values of E_1 (or e_1) and E_2 (or e_2) can be directly transferred to the primary output module, in order to control e.g. the facial expression mechanism/the mechanism simulating the endocrines of the body. (Therefore, in practice, the emotional states may be merely treated as a sort of *potentiometer.*)

10.3.3 Relationship Between the Emotion, Intention, and STM/Working Memory Modules

Apart from the aforementioned parallel functionalities of the emotion module, the module has the bi-directional connections with both the **intention** and **STM/working memory** modules as shown in Fig. 5.1. For both the connections, the connection type is essentially the same, but the amount/duration of the effect from/to these modules differs between the connection with the STM/working memory and that with the intention module:

- **Emotion $\longrightarrow$ STM/Working Memory Module**
  Sets the emotional state variables attached to the kernel unit(s) within the STM/working memory module to the current emotional states. (Or, alternatively, set the link weights between the kernel units representing the current emotional states and those within the STM/working memory.)

- **Emotion $\longrightarrow$ Intention Module**
  Gives an impact upon the states within the intention module to a certain extent, which may eventually lead to a long-term effect upon the tendency for the manner of data processing within the AMS (and thereby the overall behaviour of the body), via the intention/**thinking** module.
- **STM/Working Memory $\longrightarrow$ Emotion Module**
  Indicates the temporal (short-term) change in the emotional states, e.g. due to the memory recall from the LTM modules (and thus the activation from the corresponding kernel units) by the thinking process and/or external stimuli given to the AMS.
- **Intention $\longrightarrow$ Emotion Module**
  Gives an impact upon a relatively long-lasting tendency in the emotional states, representing *mood* or much longer *personal traits*.

Due to the relation between the emotion and intention module in the above (i.e. represented by the connections between the two modules), it is considered that the associated data processing, e.g. the memory search via the STM/working memory module, can be rather dependent upon the emotional state variables.

Related to the data processing via the aforementioned inter-module relations, it is considered that both the explicit and implicit emotional learning (for a concise review, see e.g. Gazzaniga et al., 2002) can also be interpreted within the context of the relationship between the emotion and memory modules; for both the learning, the AMS firstly receives the stimuli via the **input: sensation** module from the outside world, the binding (or data-fusion; refer back to Sect. 8.3.1) between multiple sensory data which has arrived at the STM/working memory module occurs, and the resultant network so formed is transferred to the **explicit/implicit LTM** module followed by the corresponding **primary/secondary** (i.e. perceptual) **output**. Then, the emotion module may also come into the data processing; since as in Fig. 5.1 the emotion module can be regarded as a part of the **innate structure** (as well as the sensation module), the AMS also takes into account the (emotional) state variables to a certain degree for the processing of the incoming sensory data (arrived at the STM/working memory module).

10.3.4 Implicit Emotional Learning Interpreted within the AMS Context

To be more concrete, imagine a situation where the AMS receives two different kinds of sensory data, i.e. one that can give a significant impact upon the body (or the one that does harm to the life value), whilst the other does not by itself; for instance, the pain in the wounded leg suffered in the car accident in the past (i.e. the information received as certain tactile data via the **sensation** module), which directly involves the emotion of "fear", and some sensory information of the specific car (i.e. auditory/visual) that hit the body correspond respectively to the two such different kinds of sensory data. In classical conditioning, the car and its hit to the body can be treated respectively as the conditioned stimulus (CS) and unconditioned stimulus (US), whereas the pain is an unconditioned response (UR). In the AMS context, it is considered that these two different types of sensory data were firmly bound (or associated) together and stored as a form of (at least) the two kernel units representing the respective sensory data and the link weight in between within the corresponding LTM module(s). Then, these kernel units have/share the (emotional) state variables representing the fear (i.e. by exploiting the kernel unit representation with state variables as shown in Fig. 10.3).

Next, even long after the injury is cured, such a situation is considered that once the AMS receives (only) some sort of the sensory data corresponding to the specific car (i.e. the visual sensory data corresponding to the car of the same type, such as the shape or colour, but different from the car that actually hit the body in the past), it could show a fear response, due to the retrieval of the emotional state variables (i.e. the variables attached to the respective kernel units) that can vary the current state(s) within the emotional module, the states of which can then be regarded as the conditioned response (CR), and may even follow some involuntary actions due to the activations from some other kernel units within the implicit LTM module invoked by the sensory data (i.e. due to the connections via the link weights in between). In general cognitive science/psychology, this is referred to as the *implicit emotional learning* (see e.g. Gazzaniga et al., 2002).

In addition, the duration of which such state variables within the two kernel units are so set and held can, however, be varied, during the later learning process by the AMS.

10.3.5 Explicit Emotional Learning

In contrast to the implicit emotional learning, it is possible to consider another scenario; the body was not actually involved in such an accident but acquired such knowledge of information *externally* through the relevant sensory data; i.e. imagine a situation where the AMS had captured the sensory data of the specific car (i.e. the car of the same type) and later performed the data-fusion with the fact, i.e. the information about the fact is i) received first as another sensory data, ii) processed further, and then iii) the outcome is stored within the LTM, that, e.g. the specific car had some mechanical fault and caused a traffic accident in the past. Then, similar to the previous scenario (i.e. within the context of implicit emotional learning), the AMS could vary the current emotional state by retrieving the emotional state variables (i.e. due to the memory recall during the interactive data processing amongst the associated modules) and eventually exhibit a fear response due to the functionality of the emotion module. This is in contrast referred to as the *explicit emotional learning* (see e.g. Gazzaniga et al., 2002).

10.3.6 Functionality of the Emotion Module

For both the examples of the explicit and implicit emotional learning as described above, the following conditions must, however, be met; the AMS has already acquired (i.e. due to the instinct/innateness) or learnt the fact that "one must avoid suffering from any pain for the existence of the body" and thus that "a fear is (also) associated with a pain". This is since any pain perceived can be treated as a signal that indicates a break in the body and can eventually endanger the existence.

In the AMS context, it is considered that such knowledge is pre-set within the **instinct: innate structure** module or has been learnt and stored within the **LTM** modules during the course of learning. Then, the principal role of the emotion module is to *urge* such a learning process (i.e. to initiate the memory reconfiguration process, where appropriate), in accordance with the pre-determined/stored knowledge within the instinct: innate structure and/or LTM modules (i.e. in Fig. 5.1, the links between the emotion and instinct: innate structure/LTM modules imply this functionality). In other words, the emotional states are considered as another sort of memory and thereby any single event experienced by the AMS is, in this sense, somewhat associated with the states of the body. Within the kernel memory principle, it is then considered that a single event can be eventually transformed into the template vector(s) of the kernel unit(s) (and the link weight(s) in between), whilst the emotional states are simultaneously stored within the emotional state variables attached to them (i.e. in such a case, by exploiting the modified kernel unit representation shown in Fig. 10.3).

Therefore, it is considered that the current emotional states and/or the emotional state variables attached to each kernel unit retrieved (i.e. both obtained via the **STM/working memory** and/or **intention** module) also play an essential role in the thinking process (i.e the memory search process) performed by the **thinking** module, putting aside e.g. the current condition of the link weight connections between the kernel units within the memory modules. Thereby, it is considered that the AMS can exhibit a more complicated manner of behaviours as the cause of such data processing. That is to say, the memory search process can be initiated/continued, even if the starting kernel unit does *not* have the connection with the others but holds similar emotional state variables to them. (In this sense, it is said that the memory search via the link weight connections without taking into account any emotional states is referred to as *"rational" reasoning*, in contrast to the *"emotional" reasoning*.)

In the case of the car accident example given previously (i.e. for both the explicit and implicit emotional learning cases), it is thus considered that the AMS has established a firm association (i.e. in terms of the link weights and emotional state variables) between the kernel units representing the information about the specific car and the emotional states representing the "fear", since the event is crucial to the existence of the body.

10.3.7 Stabilisation of the Internal States

In the AMS principle, the emotional states within the emotion module are always kept in such a manner that, ultimately, maximises the duration of the body, i.e. to maintain the emotional states that represent e.g. a (moderate) pleasure and relief, in accordance with the scales proposed by Rolls (Rolls, 1999), so that the entire body can maintain its balance (i.e. for the long-lasting existence of the body). This tendency can be embedded within the AMS, i.e. due to the **instinct: innate structure** module. In other words, the emotion module also functions to "suppress" excessive amount of the activities to be performed for the protection of the body. Then, in this sense, it is considered that introducing the emotion module can lead to avoidance of the so-called *frame problem* (McCarthy and Hayes, 1969; Dennett, 1984) (this notion also agrees with the philosophical standpoint. See Shibata, 2001).

In the previous car accident example, it was considered that the AMS exhibits the emotional states representing a certain level of "fear" after the implicit/explicit emotional learning of the accident event (in Sects. 10.3.4 and 10.3.5). Then, due to the innateness (i.e. the instinct: innate structure module) of the AMS, it is considered that, at a certain point, the stabilisation process starts to occur, so that the AMS resumes the emotional states representing e.g. pleasure and relief for keeping the balance of the entire body. The stabilisation process involves the associated data processing of the modules within the AMS; i.e. the **thinking** module initiates the memory search within the **LTM** (or LTM-oriented) modules and retrieves the emotional state variables from the activated kernel unit(s) within the LTM, in order to vary the current biased emotional states. This retrieval process can be facilitated further due to the functionality of the **attention** module (i.e. it is affected by way of the **intention** and/or **STM/working memory** module), since the memory search can be limited to only those which have the emotional state variables representing a "positive" emotion (or, in contrast, the current "negative" emotion can be maintained/forced, depending upon the situation).

Alternatively, such stabilisation process can, however, be omitted dependent upon the degree of the emotional learning; if the kernel network is formed as the cause of such learning process but the degree of learning to form such network is rather low, the network may eventually disappear from the memory space, or the nodes can be replaced by other kernel units (e.g. sensory data received).

10.3.8 Thinking Process to Seek the Solution to Unknown Problems

In other words, the situation where the body was involved in such an accident may also be regarded as that where the AMS encounters the problem of which a direct solution is not available.

Then, consider a situation where the AMS faces to the problem of which any solution still has yet to be found. In such a case, similar to the aforementioned memory search, the AMS resorts to a heuristic search within the LTM modules performed mainly via the thinking module, though the manner of the heuristic search may also depend heavily upon the current internal states (e.g. the emotion states) of the AMS.

10.4 Dealing with Intention

In general, the notion of "intention" can be alternatively interpreted as the aim or plan to do something[2]. In this regard, the concept of thinking is also closely tied to that of "intention", and thus it can be considered that both the concept of thinking and intention can be somewhat complementary to each other. In a similar context, the notion of "orientation" can be dealt in parallel with the "intention", though, according to the classification by Hobson (Hobson, 1999), the orientation (direction) is referred to as the spatio-temporal evocation, whilst the intention is relevant to the aim/plan.

Nevertheless, within the AMS context, the **intention** module can be regarded as the mechanism that holds temporarily the information about the *resultant states* so reached during performing the thinking process by the **thinking** module (i.e. indicated by the data flow of **thinking** $\longrightarrow$ **intention**). In reverse, the states within the intention module can to a certain extent affect the manner of the thinking process (i.e. the data flow **intention** $\longrightarrow$ **thinking**).

Then, the states so held within the intention module greatly (but indirectly) affect the memory search via the **STM/working memory** module. In terms of the temporal storage, it is thus said that the intention module also exhibits the aspect of STM/working memory (as indicated by a *dashed line*) by the parallel functionality of the intention module with the STM/working memory module in Fig. 5.1.

Within the context of kernel memory, such states can be represented by the locations/addresses of the kernel units so activated together with the emotional state variables attached to them, as well as the manner of connection(s) (i.e. represented by the kernel network(s) that consists of the kernel units so activated, where appropriate), during the thinking process. Thus, for a relatively long period of time (i.e. such a period can be varied from seconds to days or, even to years, depending upon the application/manner of implementation), the tendency in the memory search via the STM/working memory can be rather restricted to a particular type(s) of the kernel units within the LTM modules; for instance, even if the current memory search is directed to the kernel units which do not match (i.e. to a large extent) the states within the intention module (i.e. due to the focus temporally set by the **attention** or **emotion** module), once the current (or secondary) memory search is terminated (i.e. due to the thinking module, whilst sending the signals for making

---

[2]To deal with the notion "intention" (or "intentionality") in the strict philosophical sense is beyond the scope of this book.

real actions to the **primary output** module, where there are such memory accesses within the **implicit LTM** module), the primary memory search that follows the states within the intention module can be resumed.

Related to the resumption of the primary memory search due to the intention module, the small robot developed based upon the so-called "consciousness architecture" (Kitamura et al., 1995; Kitamura, 2000) can continue to perform not only the ordinary path-finding but also the chasing pursuit of another robot in a maze that is running ahead, even if e.g. it disappears from the visibility of the robot. (However, rigorously speaking, the utility of the terminology "consciousness" in their robot seems to be rather restricted in this sense; a further discussion of consciousness will be given later in Chap. 11.)

10.4.1 The Mutual Data Processing: Attention ⟷ Intention Module

As aforementioned, the intention module can also be regarded as a parallel functionality with the STM/working memory module, in that the information about the activated kernel units (and the kernel networks so formed) for a further memory search, i.e. during the thinking process performed by the thinking module, is held temporarily as the corresponding state(s). In this regard, it may be considered that the functionality is similar to the **attention** module. However, as indicated by the bi-directional data flow **intention** ⟷ **thinking** module in Fig. 5.1, the states within the intention module are directly affected by the **thinking** module and thus considered to be more oriented with the notion of *reasoning*, in comparison with the attention module. Hence, the intention module should be designed in such a way that the states within it are less susceptible to the incoming data that arrives at the STM/working memory module than the attention module.

Moreover, it is considered that the duration of keeping such information within the attention module is *shorter* than that within the intention module and hence that the functionalities of both the modules are rather complementary to each other:

- **Intention ⟶ Attention Module**
  The state(s) within the intention module normally yields the initial state(s) within the attention module, i.e. the state(s) represented in the form of the kernel network(s) e.g. during the thinking process. Then, even if the current attended kernel unit(s) is the one representing a specific domain of the data (i.e. for performing the secondary memory search) which are *not* directly relevant to the primary memory search, the aforementioned resumption of the primary memory search can take place, due to the state(s) so held within the intention module, i.e. after the completion of the secondary memory search (i.e. so judged by the thinking module) or when the memory space of the STM/working memory becomes less occupied (or in its "idle" state).

- **Attention** ⟶ **Intention Module**

  In reverse, in some situations, the attended kernel(s) (i.e. due to the attention module) can to a certain extent affect the *trend*, i.e. a relatively long tendency, of the memory search process(es) performed later/subsequently by the thinking module, by the reference to the state(s) within the intention module. For instance, the memory search can be initiated from (or limited to) the kernel unit(s) that represents a particular domain of data.

Note that, within the kernel memory principle, in contrast to the relation of the intention module with the **emotion** module (see Sect. 10.3.3), the variation in terms of the memory search process, due to the relation with the attention module, is not (primarily) dependent upon the emotional state variables but rather the link weights of the corresponding kernel units (i.e. thus relevant to the *reasoning*). Nevertheless, the manner of such implementation must be ultimately dependent upon the application; for instance, to imitate the behaviours of the real life, it is possible to design the AMS in such a way that the memory search depends more upon the emotional state variables (i.e. more aspects due to the instinct: innate structure module) than upon the interconnecting link weights.

10.5 Interpretation of Intuition

In general, *intuition* can be alternatively referred to as *instinct* or *sentience*, whilst there are other relevant notions such as *hunch*, *scent*, or the *sixth sense*. Amongst these, we here focus upon only the notion of "intuition" and how it is interpreted within the AMS context, albeit avoiding the strict sense of philosophical justification (which is beyond the scope of this book).

According to the Oxford Dictionary of English, "intuition" is the ability to understand something *instinctively* (which can also imply the close relationship between the notions of instinct and intuition, as indicated by the *dashed line* in between the two oriented modules in Fig. 5.1 (on page 84)) without the need for conscious reasoning. In contrast, as in the Japanese Dictionary (Kenbo et al., 1981), the terminology "intuition" is used to describe such a functionality based upon *experience*, whilst the relevant notion such as "hunch" is sensuous (i.e. not dependent upon any experience or reasoning) and then more closely related to the "sixth sense".

Then, as described in Sect. 8.4.6, the notion of intuition can be (partially) treated within the context of **instinct: innate structure** module and thus considered as a constituent of the (long-term) memory which holds the information regarding the physical nature of the body. In addition, it is considered that the element of learning, i.e. the aspect of *experience*, also comes in to the notion of intuition, and thus, in the AMS context, the **intuition** module must be considered within the principle of the LTM.

As in Fig. 5.1, similar to that with the aforementioned instinct module, the intuition module also has the parallel functionality with the **implicit LTM** module, since it is considered that a particular set of the data transferred via the **STM/working memory** module can activate the kernel units within the intuition module and yield the corresponding output(s) (i.e. given in the form of a series of the activations) from the **secondary output: perception** module. Thus, the intuition module also consists of multiple kernel units, as other LTM/LTM-oriented modules (in Chap. 8). Then, similar to the property of the implicit LTM module, the contents stored within such kernel units are not directly accessible from the STM/working memory module, but only the resultant perceptual outputs, i.e. given as the form of the activations from the perception module, are available. (In other words, this interpretation reflects the aforementioned notion of *understanding without the need for conscious reasoning*).

However, unlike the implicit LTM module, as indicated by the data flow **intuition** $\longrightarrow$ **thinking** in Fig. 5.1, the activations from the kernel units within the intuition module may affect directly the thinking process performed by the **thinking** module. (As described in Sect. 9.3.2, this is then somewhat relevant to the notion of *nonverbal thinking*.) Thus, in practice the degree of such affect is dependent upon implementation.

In addition, note that, in terms of the design, it is alternatively considered that the intuition module does not act as a single agent but is merely a collection of the kernel units within the implicit LTM (or other LTM-oriented) modules that may directly affect the thinking process. It is then considered that the kernel units within such a collection are chosen from those which have exhibited relatively strong activations amongst all within the LTM/LTM-oriented modules for a particular period of time (i.e. representing the experience).

So far in this chapter, we have considered the general framework of the four remaining modules within the AMS relevant to the abstract notions of mind, i.e. attention, emotion, intention, and intuition. In the forthcoming sections, we then consider how the three oriented modules, i.e. attention, emotion, and intuition module, can be actually designed within the kernel memory principle and thereby how the data processing can be performed in association with the other modules within the AMS, by examining through an example of the application for developing an intelligent pattern recognition system.

10.6 Embodiment of the Four Modules: Attention, Intuition, LTM, and STM/Working Memory Module, Designed for Pattern Recognition Tasks

In this section, we consider a practical model of a pattern recognition system by exploiting the concept of the four modules within the AMS shown in Fig. 5.1 (on page 84), i.e. attention, intuition, LTM, and STM/working memory module. In terms of the model, we will focus upon how the abstract notions related to the mind can be interpreted on a basis of an engineering framework, and thereby, we will consider how an intelligent pattern recognition system can be developed.

10.6.1 The Hierarchically Arranged Generalised Regression Neural Network (HA-GRNN) – A Practical Model of Exploiting the Four Modules: Attention, Intuition, LTM, and STM, for Pattern Recognition Systems (Hoya, 2001b, 2004b)

In recent work (Hoya, 2001b, 2004b), the author has modelled the four modules in Fig. 5.1, i.e. attention, intuition, LTM, and STM, as well as their interactive data processing, within the evolutionary process of a hierarchically arranged generalised regression neural network (HA-GRNN), the neural network of which is also proposed by the author in the literature, as shown in Fig. 10.4.

As the name HA-GRNN stands for, the model in Fig. 10.4 consists of a multiple of dynamically reconfigurable neural networks arranged in a hierarchical order, each of which can be realised by a PNN/GRNN[3] (as described in Sect. 2.3) or modified RBF-NN (i.e. for both LTM Net 1 and STM). (However, as discussed in Sect. 3.2.3, each network, i.e. for the respective LTM and STM networks, can also be regarded as the corresponding kernel memory, since PNNs/GRNNs can be subsumed into the kernel memory concept, and thus have dynamic and flexible reconfiguration properties[4].)

As depicted in Fig. 10.4, an HA-GRNN consists of a multiple of neural networks and their associated data processing mechanisms:

1) A collection of RBFs and the associated mechanism to generate the output representing the STM/LTM for yielding the "intuitive output" (denoted "LTM Net 1" in Fig. 10.4);

2) A multiple of PNNs/GRNNs representing the *regular* LTM networks (denoted "LTM Net 2-L" in Fig. 10.4);

3) A decision unit which yields the final pattern recognition result (i.e. following the so-called "winner-takes-all" strategy).

---

[3]The term HA-GRNN was preferably used, since as described in Sect. 2.3, it is considered that in practice GRNNs generalise the concept of PNNs in terms of the weight setting between the hidden and output layers.

[4]Thus, without loss of generality, within the networks of both the model in Fig. 10.4 and the extended version (which will appear in Sect. 10.7), only the RBFs (namely, Gaussian kernel functions) are considered as the respective kernel units; for the HA-GRNN, the structure of PNNs/GRNNs is considered, whereas a collection of the kernel units arranged in a matrix form is assumed for each LTM network within the extended model.

Then, both the HA-GRNN model and the extended model (to be described in Sect. 10.7) can be described within the general concept of the AMS and kernel memory principle.

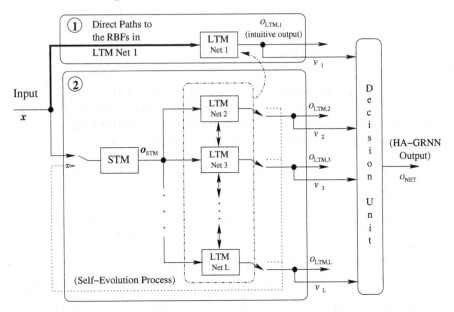

Fig. 10.4. The hierarchically arranged generalised regression neural network (HA-GRNN) – modelling the notion of attention, intuition, LTM, and STM within the evolutionary process of the HA-GRNN. As the name HA-GRNN denotes, the model consists of a multiple of dynamically reconfigurable neural networks arranged in a hierarchical order, each of which can be realised by a PNN/GRNN (see Sect. 2.3) or a collection of the RBFs and the associated mechanism to generate the output (i.e. for both LTM Net 1 and the STM)

Then, in Fig. 10.4, $\mathbf{x}$ denotes the incoming input pattern vector to the HA-GRNN, $\mathbf{o}_{STM}$ is the STM output vector, $o_{LTM,i}$ $(i = 1, 2, \ldots, L)$ are the LTM network outputs, v_i are the respective weighting values for the LTM network outputs, and o_{NET} is the final output obtained from the HA-GRNN (i.e. given as the pattern recognition result by 3) above).

The original concept of the HA-GRNN was motivated from various studies relevant to the memory system in the brain (James, 1890; Hikosaka et al., 1996; Shigematsu et al., 1996; Osaka, 1997; Taylor et al., 2000; Gazzaniga et al., 2002).

10.6.2 Architectures of the STM/LTM Networks

As in Fig. 10.4, the LTM networks are subdivided into two types of networks; one for generating "intuitive outputs" ("LTM Net 1") and the rest ("LTM Net 2 to LTM Net L") for the regular outputs.

For the regular LTM, each LTM Net (2 to L) is the original PNN/GRNN (and thus has the same structure as shown in the right part of Fig. 2.2, on

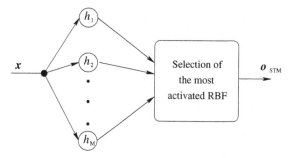

Fig. 10.5. The architecture of the STM network – consisting of multiple RBFs and the associated LIFO stack-like mechanism to yield the network output. Note that the STM network output is given as a *vector* instead of a scalar value

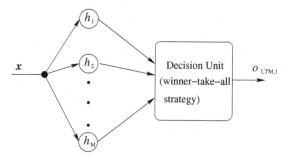

Fig. 10.6. The architecture of LTM Net 1 – consisting of multiple RBFs and the associated mechanism to yield the network output (i.e. by following the "winner-takes-all" strategy)

page 15), whereas both the STM and LTM Net 1 consist of a set of RBFs and the associated mechanism to generate the output from the network (alternatively, they can also be seen as modified RBF-NNs) as illustrated in Figs. 10.5 and 10.6, respectively. As described later, the manner of generating outputs from STM or LTM Net 1 is however different from ordinary PNNs/GRNNs.

Although both the architectures of the STM and LTM Net 1 are similar to each other, the difference is left within the manner of yielding the network output; unlike ordinary neural network principle, the network output of the STM is given as the *vector* obtained by the associated LIFO stack-like mechanism (to be described later in Sect. 10.6.4), whilst that given by LTM Net 1 is a scalar value as in ordinary PNNs/GRNNs.

10.6.3 Evolution of the HA-GRNN

The HA-GRNN is constructed by following the *evolutionary* schedule which can be subdivided further into the following five phases:

[Evolutionary Schedule of HA-GRNN]

Phase 1: The STM and LTM Net 2 formation.
Phase 2: Formation/network growing of LTM Nets (2 to L).
Phase 3: Reconfiguration of LTM Nets (2 to L) (self-evolution).
Phase 4: Formation of LTM Net 1 (for generating intuitive outputs).
Phase 5: Formation of the attentive states.

Phase 1: Formation of the STM Network and LTM Net 2

In Phase 1, the STM network is firstly formed (how the STM network is actually formed will be described in detail in Sect. 10.6.4), and then LTM Net 2 is constructed by directly assigning the output vectors of the STM network to the centroid vectors of the RBFs in LTM Net 2. In other words, at the initial stage of the evolutionary process (i.e. from the very first presentation of the incoming input pattern vector until LTM Net 2 is filled), since each LTM network except LTM Net 1 is represented by a PNN/GRNN, the RBFs within LTM Net 2 are distributed into the respective sub-networks, according to the class "label" (i.e. the label is set by the target vector consisting of a series of indicator functions as defined in (2.4); cf. also Fig. 2.2, on page 15) associated with each centroid vector.

Phase 2: Formation of LTM Nets (2 to L)

The addition of the RBFs in Sub-Net i ($i = 1, 2, \ldots, N_{cl}$, where N_{cl} is the number of classes which is identical to the number of the sub-nets in each LTM network[5]) of LTM Net 2 is repeated until the total number of RBFs in Sub-Net i reaches a maximum $M_{LTM_2,i}$ (i.e. the process can be viewed as the network growing). Otherwise, the least activated RBF in Sub-Net i is moved to LTM Net 3. Then, this process corresponds to Phase 2 and is summarised as follows:

[Phase 2: Formation of LTM Nets (2 to L)]

Step 1)

Provided that the output vector from the STM network falls into Class i, for $j = 1$ to $L-1$, perform the following:

If the number of the RBFs in Sub-Net i of LTM Net j reaches a maximum $M_{LTM_{j,i}}$, move the least activated RBF within Sub-Net i of LTM Net j to that of LTM Net $j + 1$.

---

[5]Here, without loss of generality, it is assumed that the number of the sub-nets is unique in each of LTM Nets (2 to L).

Step 2)
>   If the number of the RBFs in Sub-Net i of LTM
>   Net L reaches a maximum $M_{LTM_{L,i}}$ (i.e. all the i-th
>   sub-networks within LTM Nets (2 to L) are filled), there
>   is no entry to store the new output vector. Therefore,
>   perform the following:
>
>   **Step 2.1)** Discard the least activated RBF in Sub-Net
> i of LTM Net L.
>   **Step 2.2)** Shift one by one all the least activated RBFs
>   in Sub-Net i of LTM Nets (L-1 to 2) into that of
>   LTM Nets (L to 3).
>   **Step 2.3)** Then, store the new output vector from the
>   STM network in Sub-Net i of LTM Net 2.
>   (Thus, it can be seen that the procedure above is
>   also similar to a last-in-first-out (LIFO) stack; cf.
>   the similar strategy for the STM/working memory
>   module described in Sect. 8.3.7.)

The above process is performed based on the hypothesis that long-term memory can be represented by a layered structure, where in the HA-GRNN context the (regular) long-term memory is represented as a group of LTM Nets (2 to L), and that each element of memory is represented by the corresponding RBF and stored in a specific order arranged according to the contribution to yield the final output of the HA-GRNN.

In Fig. 10.4, the final output from the HA-GRNN o_{NET} is given as the largest value amongst the weighted LTM network outputs $o_{LTM,i}$ ($i = 1, 2, \cdots, L$):

$$o_{NET} = \max(v_1 \times o_{LTM,1}, v_2 \times o_{LTM,2}, \ldots, v_L \times o_{LTM,L}), \quad (10.3)$$

where

$$v_1 \gg v_2 > v_3 > \ldots > v_L. \quad (10.4)$$

Note that the weight value v_1 for $o_{LTM,1}$ must be given relatively larger than the others $v_2, v_3, \ldots, v_L$. This discrimination then urges the formation of the *intuitive output* from the HA-GRNN to be described later.

Phase 3: Reconfiguration of LTM Nets (2 to L) (Self-Evolution)

After the formation of LTM Nets (2 to L), the reconfiguration process of the LTM networks may be initiated in Phase 3, in order to restructure the LTM part. This process may be invoked either at a particular (period of) time or due to the strong excitation of some RBFs in the LTM networks by a particular input pattern vector(s)[6]. During the reconfiguration phase, the presentation of the incoming input pattern vectors from the outside is not allowed to process at all, but the centroid vectors obtained from the LTM networks are used instead as the input vectors to the STM network (hence the term "self-evolution"). Then, the reconfiguration procedure within the HA-GRNN context is summarised as follows:

[Phase 3: Reconfiguration of LTM Nets (2 to L)
(Self-Evolution)]

Step 1)
    Collect all the centroid vectors within LTM Nets 2 to l
    ($l \leq L$), then set them as the respective incoming pattern
    vectors to the HA-GRNN.
Step 2)
    Present them to the HA-GRNN, one by one. This process
    is repeated p times. (In Fig. 10.4, this flow is depicted
    (*dotted line*) from the regular LTM networks to the STM
    network.)

It is then considered that the above reconfiguration process invoked at a particular time period is effective for "shaping up" the pattern space spanned by the RBFs within LTM Nets (2 to L).

In addition, alternative to the above, such a non-hierarchical clustering method as in (Hoya and Chambers, 2001a) may be considered for the reconfiguration of the LTM networks. The approach in (Hoya and Chambers, 2001a) is, however, not considered to be suitable for the instance-based (or rather hierarchical clustering) operation as above, since, with the approach in (Hoya and Chambers, 2001a), a new set of the RBFs for LTM will be obtained by *compressing* the existing LTM using a clustering technique, which, as reported, may (sometimes) eventually collapse the pattern space, especially when the number of representative vectors becomes small.

Phase 4: Formation of LTM Net 1

In Phase 4, a certain number of the RBFs in LTM Nets (2 to L) which keep relatively strong activation in a certain period of the pattern presentation are transferred to LTM Net 1. Each RBF newly added in LTM Net 1 then forms a modified PNN/GRNN and will have a direct connection with the incoming input vector, instead of the output vector from the STM. The formation of LTM Net 1 is summarised as follows[7]:

---

[6]In the simulation example given later, the latter case will not be considered due to the analytical difficulty.

[7]Here, although the LTM is divided into the regular LTM networks (i.e. LTM Nets 2 to L) and LTM Net 1 for generating the intuitive outputs, such a division

---

[Phase 4: Formation of LTM Net 1]

Step 1)

In Phases 2 and 3 (i.e. during the formation/reconfiguration of the LTM Nets (2 to L)), given an output vector from the STM, the most activated RBFs in LTM Nets (2 to L) are monitored; each RBF has an auxiliary variable which is initially set to 0 and is incremented, whenever the corresponding RBF is most activated *and* the class ID of the given incoming pattern vector matches the sub-network number to which the RBF belongs.

Step 2)

Then, at a particular time or period (q, say), list up all the auxiliary variables (or, activation counter) of the RBFs in LTM Nets (2 to L) and obtain the N RBFs with the N largest numbers, where the number N can be set as

$N << \sum_i \sum_j M_{LTM_{j,i}}$ $(j = 2, 3, ..., L)$.

Step 3)

If the total number of RBFs in LTM Net 1 is currently less than or equal to $M_{LTM_1} - N$ (i.e. M_{LTM_1} denotes the maximum number of the RBFs in LTM Net 1, assuming $N \leq M_{LTM_1}$), move all the N RBFs to LTM Net 1. Otherwise, retain the original $M_{LTM_1} - N$ RBFs within LTM Net 1 and fill/replace the remaining RBFs in LTM Net 1 with the N newly obtained RBFs.

Step 4)

Create a direct path to the incoming input pattern vector for each RBF added in the previous step[8]. (This data flow is illustrated (*bold line*) in Fig. 10.4.) The output of LTM Net 1 is given as a maximum value within all the activations of the RBFs (i.e. calculated by (3.13) and (3.17)).

Note that, unlike other LTM networks, the radii values of the RBFs in LTM Net 1 must *not* be varied during the evolution, since the strong activation

---

may not be actually necessary in implementation; it is considered that the input vectors to some of the RBFs within the LTM networks are simply changed from o_{STM} to x. Then, the collection of such RBFs represents LTM Net 1.

[8]In the HA-GRNN shown in Fig. 10.4, the LTM Net 1 corresponds to the intuition module within the AMS context. However, as shown in the figure, a direct path is created to each RBF without passing through the STM network (i.e. corresponding to the STM/working memory module). This is since the STM network in the HA-GRNN is designed so that it *always* perform the buffering process to be described later. However, here the general concept of the STM/working memory module within the AMS context is still valid in the sense that the intuitive outputs can be quickly generated without a further data processing within the STM.

from each RBF (for a particular set of pattern data) is expected to continue after the transfer with the current radii values.

Up to here, the first four phases within the evolutionary process of HA-GRNN have been described in detail. Before moving on to the discussion of how the process in Phase 4 above can be interpreted as the notion of intuition and the remaining Phase 5, the latter of which is relevant to the other notion, attention, we next consider the associated data processing within the STM network in more detail.

10.6.4 Mechanism of the STM Network

As depicted in Fig. 10.5, the STM network consists of multiple RBFs and the associated mechanism to yield the network output, which selects the maximally activated RBF (centroid) and then passes the centroid vector as the STM network output. (Thus, the manner of generating the STM network outputs differs from those of LTM Nets 1-L.) Unlike LTM Nets 1-L, the STM network itself is *not* a pattern classifier but rather functions as a sort of *buffering/filtering* process of the incoming data by choosing a maximally activated RBF amongst the RBFs present in the STM, imitating the functionality of e.g. the hippocampus in the real brain to store the data within the LTM (see Sect. 8.3.2). Then, it can be seen that the output from the STM network is given as the *filtered* version of the incoming input vector $\mathbf{x}$.

Note also that, unlike the regular LTM networks (i.e. LTM Nets 2-L), the STM network does not have any sub-networks of its own; it is essentially based upon a single layered structure which is comprised by a collection of RBFs, where the maximum number of RBFs is fixed to M_{STM}. (Then, the number M_{STM} represents the *memory capacity* of the STM.) Thus, as LTM Nets (2-L) described earlier, the STM is also equipped with a mechanism similar to a last-in-first-out (LIFO) stack queue due to the introduction of the factor M_{STM}.

The mechanism of the STM network is then summarised as follows:

[Mechanism of the STM Network]

Step 1)
- If the number of RBFs within the STM network $M < M_{STM}$, add an RBF with activation h_i (i.e. calculated by (2.3)) and its centroid vector $\mathbf{c}_i = \mathbf{x}$ in the STM network. Then, set the STM network output vector $\mathbf{o}_{STM} = \mathbf{x}$. Terminate.
- Otherwise, go to Step 2).

Step 2)

- If the activation of the least activated RBF (h_j, say) $h_j < \theta_{STM}$, replace it with a new one with the centroid vector $\mathbf{c}_j = \mathbf{x}$. In such a case, set the STM network output $\mathbf{o}_{STM} = \mathbf{x}$.
- Otherwise, the network output vector $\mathbf{o}_{STM}$ is given as the *filtered* version of the input vector $\mathbf{x}$, i.e:

$$\mathbf{o}_{STM} = \lambda\mathbf{c}_k + (1-\lambda)\mathbf{x} \qquad (10.5)$$

where $\mathbf{c}_k$ is the centroid vector of the most activated RBF (k-th, say) h_k within the STM network and λ is a *smoothing* factor ($0 \le \lambda \le 1$).

In Step 2) above, the smoothing factor λ is introduced in order to determine how fast the STM network is evolved by a new instance (i.e. the new incoming pattern vector) given to the STM network. In other words, the role of this factor is to determine how quickly the STM network is responsive to the new incoming pattern vector and *switches* its focus to the patterns in other domains. Thus, this may somewhat pertain to the *selective attention* of a particular object/event. For instance, if the factor is set small, the output $\mathbf{o}_{STM}$ becomes more likely to the input vector $\mathbf{x}$ itself. Then, it is considered that this imitates the situation of "carelessness" by the system. In contrast, if the factor is set large, the STM network can "cling" to only a particular domain set of pattern data. Then, it is considered that the introduction of this mechanism can contribute to the attentional functionality within the HA-GRNN to be described in Sect. 10.6.6.

10.6.5 A Model of Intuition by an HA-GRNN

In Sect. 10.5, it was described that the notion of intuition can be dealt within the context of *experience* and is thus considered that the intuition module can be designed within the framework of LTM.

Based upon this principle, another form of LTM network, i.e. LTM Net 1, is considered within the HA-GRNN; in Fig. 10.4, there are two paths for the incoming pattern vector $\mathbf{x}$, and, unlike regular LTM networks (i.e. LTM Nets 2-L), the input vector $\mathbf{x}$ is directly transferred to LTM Net 1 (apart from the STM network), whilst, in Fig. 5.1, the input data are given to the **intuition** module via the **STM/working memory** module. Within the AMS context, this formation corresponds to the possible situation where, the input data transferred via the STM/working memory module can also activate some of the kernel units within the intuition module, whilst the input data (temporarily) stay within the STM/working memory module.

Then, the following conjecture can be drawn:

> **Conjecture 1**: In the context of HA-GRNN, the notion of *intuition* can be interpreted in such a way that, for the incoming input pattern vectors that fall in a particular domain, there exists a certain set of the RBFs that keep relatively strong activation amongst all the RBFs within the LTM networks.

The point of having these two paths within the HA-GRNN is therefore that for the *regular* incoming pattern data the final output will be generated after the associated processing within the two-stage memory, namely the STM and LTM, whilst a certain set of input patterns may excite the RBFs within LTM Net 1, which is enough to yield the "intuitive" outputs from the HA-GRNN. Then, the evidence for referring to the output of LTM Net 1 as intuitive output is that, as in the description of the evolution of HA-GRNN in Sect. 10.6.3, LTM Net 1 will be formed *after a relatively long and iterative exposition* of incoming pattern vectors, which results in the strong excitation of (a certain number of) the RBFs in LTM Nets (2 to L). In other words, the transition of the RBFs from the STM to LTM Nets (2 to L) corresponds to a *regular* learning process, whereas, in counter-wise, that from LTM Nets (2 to L) to LTM Net 1 gives the chances of yielding the "intuitive" outputs from the HA-GRNN. (Therefore, the former data flow, i.e. the STM network $\longrightarrow$ LTM Nets (2 to L) thus corresponds to the data flow **STM/working memory** $\longrightarrow$ **LTM** modules, whereas the latter indicates the reconfiguration of the LTM, implied by the relationship between the LTM and intuition modules within the AMS context; see Sects. 8.3.2 and 10.5.)

In practice, this feature is particularly useful, since it is highly expected that the HA-GRNN can generate faster and simultaneously better pattern recognition results from LTM Net 1, whilst keeping the entire network size smaller than e.g. the conventional MLP-NN trained by an iterative algorithm (such as BP) with a large amount of (or whole) training data, than the ordinary reasoning process, i.e. the reasoning process through the STM + regular LTM Nets (2 to L).

In contrast, we quite often hear such episodes as, "I have got a flash to a brilliant idea!" or "Whilst I was asleep, I was suddenly awaken by a horrible nightmare." It can also be postulated that all these phenomena occur in the brain, similar to the data processing of intuition, during the self-evolution process of memory. Within the context of HA-GRNN, this is relevant to Phase 3 in which, during the reconfiguration (or, reconstruction, in other words) phase of the LTM, some of the RBFs in LTM are excited enough to exceed a certain level of activation. Then, these RBFs remain in LTM for a relatively long period, or even (almost) perpetually, because of such memorable events to the system (therefore this is also somewhat related to the explicit/implicit emotional learning; see Sects. 10.3.4 and 10.3.5).

Moreover, it is said that this interpretation is also somewhat relevant to the psychological justifications (Hovland, 1951; Kolers, 1976), in which the authors state that, once one has acquired the behavioral skill (i.e. the notion is relevant to procedural memory), the person would not forget it for a long time. Therefore, this view can also support the notion of the parallel functionality of the intuition module with the **implicit LTM** module (as implicitly shown in Fig. 5.1, on page 84).

10.6.6 Interpreting the Notion of Attention by an HA-GRNN

Within the HA-GRNN context, the notion of attention is to focus the HA-GRNN on a particular set of incoming patterns, e.g. imitating the situation of paying attention to someone's voice or the facial image, in order to acquire further information of interest, in parallel to process other incoming patterns received by the HA-GRNN, and, as described in Sect. 10.6.4, the STM network has the role.

Phase 5: Formation of Attentive States

In the model of maze-path finding (Kitamura et al., 1995; Kitamura, 2000), the movement of the artificial mouse is controlled by a mechanism, i.e. the so-called "consciousness architecture"[9], in order to continue the path-finding pursuit, by the introduction of a higher layer of memory representing the state of "being aware" of the path-finding pursuit, whilst the lower part is used for the actual movement. Then, it is said that the model in (Kitamura et al., 1995; Kitamura, 2000) exploits a sort of "hierarchical" structure representing the notion of attention.

In contrast, within the HA-GRNN context, another hierarchy can be represented by the number of RBFs within the STM network:

> **Conjecture 2**: In the HA-GRNN context, the state of being "attentive" of something is represented in terms of a particular set of RBFs within the STM network.

Then, it is said that the conjecture above (moderately) agrees with the notion of attention within the AMS context, in that a particular subset of kernel units within the STM/working memory module contribute to the associated data processing due to the attention module (refer back to Sect. 10.2.1). (In addition, the conjecture above is also relevant to the data flow **attention** $\longrightarrow$ **STM/working memory module** within the AMS.) In the HA-GRNN, the attentive states can then be formulated during Phase 5:

---

[9]Strictly, the utility of the term "awareness" seems to be more appropriate in the context.

[Phase 5: Formation of Attentive States]

Step 1)

Collect $m(\leq M_{STM})$ RBFs of which the auxiliary variables are the first m largest amongst all the RBFs within LTM Nets (1-L), for given particular classes. Each auxiliary variable is a counter that is attached to the corresponding RBF and reports the number of excitations from. (In terms of the kernel memory, the variable corresponds to the excitation counter ε, i.e. cf. Fig. 3.1, 3.2, or 10.3.) Then, such a collection forms the attentive states of the HA-GRNN.

Step 2)

Add the copies of the m RBFs back into the STM network, whilst the $M_{STM} - m$ most activated RBFs in the STM network remain intact. The m RBFs so chosen remain within the STM for a certain long period, without updating their centroid vectors (whereas the radii may be updated).

In the above, it may also be viewed that the data flow of **LTM modules** $\longrightarrow$ **STM/working memory module** within the AMS is realised by the selection process of the RBFs (or generally kernel units) and then copying them back to the STM network (cf. the memory recall process for the data-fusion in Sect. 8.3.2). Moreover, it is said that this is in contrast to the regular learning process (i.e. refer back to Sect. 10.6.5), i.e. the data flow: the STM network $\longrightarrow$ LTM Net 2-L.

Then, in Phase 5, the m RBFs so selected make the HA-GRNN focus upon a particular (domain) set of incoming input vectors, and, by increasing m, it is expected that the filtering process in transferring incoming pattern vectors to the LTM networks becomes more accurate for particular classes. For instance, if the HA-GRNN is applied to pattern recognition tasks, it is expected that the system can compensate for the misclassified patterns that fall in to a certain class(es). In addition, the radii values of the m RBFs so copied may be updated in due course, since the parameters of the other remaining RBFs within the STM network can be varied during the course of learning.

Therefore, it is postulated that the ratio between the m RBFs and the rest of the $M_{STM} - m$ RBFs in the STM networks determines the "level" of attention. Thereby, the following conjecture can also be drawn:

Conjecture 3: The level of attention can be determined by the ratio between the number of m most activated RBFs selected from the LTM networks and that of the remaining $M_{STM} - m$ RBFs within the STM network.

Thus, Conjecture 3 also suggests that, as in the Baddeley & Hitch's working memory (in Sect. 8.3.1), the level of attention can to a large extent affect the consolidation of the LTM during the rehearsal process within the STM/working memory; in the context of an HA-GRNN, an incoming pattern vector (or a set of the input pattern vectors) can be compared to the input information to the brain and is temporarily stored within the STM network (hence the function of *filtering* or *buffering*). Then, during the evolution, the information represented by the RBFs within the STM network is selectively transferred to the LTM networks, as in Phases 1–3. In contrast, the RBFs within the LTM networks may be transferred back to the STM, because the "attention" of certain classes (or those RBFs) occurs at particular moments. (This interaction can also be compared to the "learning" process in Hikosaka et al. (1996).)

Unlike the AMS, in the original HA-GRNN context, since the evolution process is, strictly speaking, not autonomous, we may want to pre-set the state of the "attention" in advance, according to the problems encountered in practical situations. (However, it is still possible to evolve the HA-GRNN autonomously by appropriately setting the transition operations suited for a specific application, though such a case is not considered here.) For instance, in the context of pattern recognition tasks, one may limit the number of the classes to $N < N_{cl}$ in such a way that "For a certain period of the pattern presentations, the HA-GRNN must be attentive to only N classes amongst a total of N_{cl}", in order to *reinforce* the performance of the HA-GRNN for the particular N classes."

10.6.7 Simulation Example

Here, we consider a simulation example of the HA-GRNN applied to the pattern recognition tasks using the data sets extracted from the three databases, i.e. the SFS (Huckvale, 1996), OptDigit, and PenDigit database (for the description of the three databases, see also Sect. 2.3.5).

In the simulation, the data set for the SFS consisted of a total of 900 utterances of the digits from /ZERO/ to /NINE/ by nine different English speakers (including both the female and male speakers). The data set was then arbitrarily partitioned into two sets; one for constructing an HA-GRNN (i.e. the incoming pattern/training set) and the other for testing (i.e. unknown to the HA-GRNN). The incoming pattern set contains a total of 540 feature patterns, where 54 patterns were chosen for each digit, whilst the testing consists of a total of 360 patterns (i.e. 36 per digit). In both the sets, each pattern was comprised of a feature vector with a normalised set of 256 data points obtained by applying the same LPC-Mel-Cepstral analysis (Furui, 1981) as the one in Sect. 2.3.5. The feature vector was thus used as an input pattern vector to the HA-GRNN **x**.

Table 10.1. Network configuration parameters for the HA-GRNN used in the simulation example

Parameter	SFS	OptDigit	PenDigit
Max. num. of centroids in STM, M_{STM}	30	30	30
Total num. of LTM networks, $(L+1)$	3	2	4
Max. num. of centroids in LTM Net 1, M_{LTM_1}	5	25	15
Num. of sub-networks in LTM Nets 2-L, N_{cl}	10	10	10
Max. num. of centroids in each subnet, $M_{LTM_{j,i}}$ $(j = 2, 3, \ldots, L, i = 1, 2, \cdots, 10)$	4	2	4

In contrast, both the OptDigit and PenDigit data sets were composed of 1200 and 400 feature vectors for the construction and testing sets, respectively. As summarised in Table 2.1, each of the feature vectors has 64 data points for the OptDigit, whereas 16 data points for the PenDigit.

Parameter Setting of the HA-GRNN

In Table 10.1, the network configuration parameters of the HA-GRNN used in the simulation example are summarised. In the table, $M_{LTM_1}, M_{LTM_{2,i}}$, and $M_{LTM_{3,i}}$ (i.e. for the SFS; $i = 1, 2, \ldots, 10$, corresponding to the respective class IDs, $1, 2, \ldots, 10$) were arbitrarily chosen, whilst N_{cl} was fixed to the number of the classes (i.e. the ten digits). With this setting, the total number of RBFs in LTM Nets (1 to 3, for the SFS), $M_{LTM,Total}$ is thus calculated as

$$M_{LTM,Total} = M_{LTM,1} + N_{cl}(M_{LTM,2} + M_{LTM,3})$$

which yields i) 85 for the SFS, ii) 65 for the OptDigit, and iii) 175 for the PenDigit data set, respectively.

The STM Network Setting

For the STM network, both the choices of M_{STM} (as shown in Table 10.1) and the unique radius setting $\theta_\sigma = 2$ in (2.6) were made *a priori* so that the STM network functions as a "buffer" to the LTM networks with sparsely but reasonably covering all the ten classes during the evolution. Then, the setting of $\theta_{STM} = 0.1$ (i.e the threshold value of the activation of the RBFs in the STM network) and the smoothing factor $\lambda = 0.6$ in (10.5) were used for all the three data sets. (In the preliminary simulation, it was empirically found that the choice of $\lambda = 0.6$ yields a reasonable generalisation performance of the HA-GRNN.)

Parameter Setting of the Regular LTM Networks

For the radii setting of LTM Nets (2 to L), the unique setting of $\theta_\sigma = 0.25$ for both the SFS and OptDigit or $\theta_\sigma = 0.05$ in (2.6) for the PenDigit was empirically found to be a choice for maintaining a reasonably good generalisation

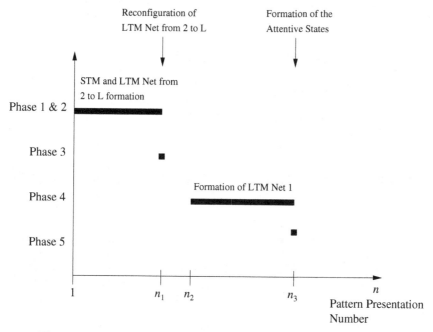

Fig. 10.7. The evolution schedule used for the simulation example

capability during the evolution. Then, to give the "intuitive" outputs from LTM Net 1, the weighting factor v_1 was fixed to 2.0, whilst the remaining v_i ($i = 2, 3, \ldots, L$) were given by the linear decay

$$v_i = 0.8(1 - 0.05(i - 2)) .$$

The Evolution Schedule

Figure 10.7 shows the evolution schedule used for the simulation example. In the figure, the index n corresponds to the presentation of the n-th incoming pattern vector to the HA-GRNN. In the simulation, the setting $n_2 = n_1 + 1$ was used, without loss of generality. Note that the formation of LTM Net 1 was scheduled to occur after a relatively long exposition of incoming input vectors (thus $n_1 < n_2$), as described in Sect. 10.6.5. Then, note that, with this setting, it requires that the RBFs in LTM Net 1 should be effectively selected from the previously (i.e. the time before n_1) spanned pattern space in the LTM networks. Thus, the self-evolution (in Phase 3) was scheduled to occur at n_1 with $p = 2$ in the simulation (i.e. the self-evolution was performed twice at $n = n_1$, and it was empirically found that this setting does not give any impact upon the generalisation performance).

Table 10.2 summarises the setting of n_1 and n_3 (which covers all the five phases) used for the simulation example. Then, the evolution was eventually

Table 10.2. Parameters for the evolution of the HA-GRNN used for the simulation example

Parameter	SFS	OptDigit	PenDigit
n_1	200	400	400
n_3	400	800	800

Table 10.3. Confusion matrix obtained by the HA-GRNN after the evolution – using the SFS data set

Digit	0	1	2	3	4	5	6	7	8	9	Total	Generalisation Performance
0	29				3		2	1		1	29/36	80.6%
1		31			1	2				2	31/36	86.1%
2	1		28	2	2		1	2			28/36	77.8%
3				32	2	1		1			32/36	88.9%
4					36						36/36	100.0%
5		3			1	27		2		3	27/36	75.0%
6							32	2	2		32/36	88.9%
7								36			36/36	100.0%
8						1	1	34			34/36	94.4%
9		4			10			1		21	21/36	58.3%
Total											306/360	85.0%

stopped when all the incoming pattern vectors in the training set were presented to the HA-GRNN.

Simulation Results

To evaluate the overall recognition capability of the HA-GRNN, all the testing patterns were presented one by one to the HA-GRNN, and the generalisation performance over the testing set was obtained after the evolution from the decision unit (i.e. given as the final HA-GRNN output o_{NET} in Fig. 10.4). For the intuitive outputs, the generalisation performance obtained from LTM Net 1 during testing was also considered.

Table 10.3 shows the confusion matrix obtained by the HA-GRNN after the evolution using the SFS data set. In this case, no attentive states were considered at n_3.

For comparison of the generalisation capability, Table 10.4 shows the confusion matrix obtained using a conventional PNN with the same number of RBFs in each subnet (see Fig. 2.2 on page 15) as the HA-GRNN (i.e. a total of 85 RBFs were used), where the respective RBFs were found by the well-known MacQueen's k-means clustering method (MacQueen, 1967). To give a fair comparison, the RBFs in each subnet were obtained by applying the k-means clustering to the respective (incoming pattern vector) subsets containing 54 samples per each digit (i.e. from Digit /ZERO/ to /NINE/).

Table 10.4. Confusion matrix obtained by the conventional PNN using k-means clustering method – using the SFS data set

Digit	0	1	2	3	4	5	6	7	8	9	Total	Generalisation Performance
0	34		1		1						34/36	94.4%
1		17			19						17/36	47.2%
2			28		8						28/36	77.8%
3			3	22	10	1					22/36	61.1%
4					36						36/36	100.0%
5						36					36/36	100.0%
6							36				36/36	100.0%
7	1		3		2	5	6	19			19/36	52.8%
8					2	1	7		26		26/36	72.2%
9			1		27					8	8/36	22.2%
Total											262/360	72.8%

In comparison with the conventional PNN as in Table 10.4, it is evidently observed in Table 10.3 that, besides the superiority in the overall generalisation capability of the HA-GRNN, the generalisation performance in each digit (except Digit /NINE/) is relatively consistent, whilst the performance with the conventional PNN varies dramatically from digit to digit as in Table 10.4. This indicates that the pattern space spanned by the RBFs obtained using the k-means clustering method is rather biased.

Generation of the Intuitive Outputs

For the SFS data set, the intuitive outputs were generated three times during the evolution, and all the three patterns were correctly classified for Digits /FOUR/ and /EIGHT/. In contrast, during testing, 13 pattern vectors amongst 360 yielded the generation of the intuitive outputs from LTM Net 1 in which 12 out of the 13 patterns were correctly classified. It was then observed that the Euclidean distances between the twelve pattern vectors and the respective centroid vectors corresponding to their class IDs (i.e. digit numbers) were relatively small and, for some patterns, close to the minimum (i.e. the distance between that of Pattern Nos. 77, 88, 104, and 113, and the RBFs for Digits /SEVEN/, /EIGHT/, /FOUR/, and /THREE/, respectively, in LTM Net 1 were minimal). From this observation, it can therefore be confirmed that, since intuitive outputs are likely to be generated when the incoming pattern vectors are rather close to the respective centroid vectors in LTM Net 1, the centroid vectors correspond to the notion of "experience".

For the OptDigit, despite the slightly worse generalisation capability by HA-GRNN (87.0%) compared with that of the PNN with k-means (88.8%), the generalisation performance for the 174 out of the 360 testing patterns which yielded the intuitive outputs was better, i.e. 95.1%. This indicates that the LTM Net 1 was successfully formed and contributed to the improved performance. Moreover, as discussed in Sect. 10.6.5, this leads to a faster decision-making, since the intuitive outputs were generated, e.g. without the processing within the STM network and the regular LTM Nets.

In contrast, for the PenDigit, whilst overall a better generalisation performance was obtained by the HA-GRNN (89.3%) in comparison with that of the conventional PNN (88.0%), only a single testing pattern yielded the intuitive output (in which the pattern was correctly classified). Then, by increasing the maximum number of allowable RBFs in LTM Net 1 (as in Table 10.1, which was initially fixed to 15), to 100, the simulation was performed again. As expected, the number of times that intuitive outputs are generated was increased to 14, in which all the 14 testing patterns were correctly classified.

Simulations on Modelling the Attentive States

In Table 10.3, it is observed that the generalisation performance for Digits /FIVE/ and /NINE/ is relatively poor. To study the effectiveness of having the attentive states within the HA-GRNN, the attentive states were considered for both Digits /FIVE/ and /NINE/.

Then, by following both the conjectures 2 and 3 in Sect. 10.6.6, 10 (20 for the PenDigit) amongst a total of 30 RBFs within the STM network were fixed for the respective digits after evolution time n_3. In addition, since the poor generalisation performance for Digits /FIVE/ and /NINE/ was (perhaps) due to the insufficient number of the RBFs within LTM Nets (2 to 3), the maximum number $M_{LTM_{2,i}}$ and $M_{LTM_{3,i}}$ ($i = 5$ and 10), respectively, were also increased.

Table 10.5 shows the confusion matrix obtained by the HA-GRNN configured with an attentive state of only Digit /NINE/. For this case, a total of 8 more RBFs in LTM Nets 2 and 3 (i.e. 4 more each in LTM Nets 2 and 3) which correspond to the first 8 (instead of 4) strongest activations were

Table 10.5. Confusion matrix obtained by the HA-GRNN after the evolution – with an attentive state of Digit 9 – using the SFS data set

Digit	0	1	2	3	4	5	6	7	8	9	Total	Generalisation Performance
0	29			1	3		2	1			29/36	80.6%
1		31			2	2				1	31/36	86.1%
2	1		28	2	2		1	2			28/36	77.8%
3				32	2	1		1			32/36	88.9%
4					36						36/36	100.0%
5		2			1	29		2		2	29/36	80.6%
6							32	2	2		32/36	88.9%
7								36			36/36	100.0%
8						1		1	34		34/36	94.4%
9		2			11					23	23/36	63.9%
Total											310/360	86.1%

Table 10.6. Confusion matrix obtained by the HA-GRNN after the evolution – with an attentive state of Digits 5 and 9 – using the SFS data set

Digit	0	1	2	3	4	5	6	7	8	9	Total	Generalisation Performance
0	29			1	3		2				29/36	80.6%
1		31		2	2					1	31/36	86.1%
2	1		28	2	2	1	2				28/36	77.8%
3				33	2		1				33/36	91.7%
4					36						36/36	100.0%
5		1			1	33				1	33/36	91.7%
6							32	2	2		32/36	88.9%
7			4					36			36/36	100.0%
8							1	1	34		34/36	94.4%
9		3	1		8					24	24/36	66.7%
Total											316/360	87.8%

selected (following Phase 2 in Sect. 10.6.3) and added into Sub-Net 10 within both the LTM Nets 2 and 3 (i.e. accordingly, the total number of RBFs in LTM Nets (1 to 3) was increased to 93). As in the table, the generalisation performance of Digit /NINE/ was improved at 63.9%, in comparison with that in Table 10.3, whilst preserving the same generalisation performance for other digits.

In contrast, Table 10.6 shows the confusion matrix obtained with having the attentive states of both the digits /FIVE/ and /NINE/. Similar to the case with a single attentive state of Digit /NINE/, a total of 16 such RBFs for the two digits were respectively added into Sub-Nets 6 and 10 within both the LTM Nets 2 and 3. (Thus, the total number of RBFs in LTM Nets (1 to 3) was increased to 101.) In comparison with Table 10.3, the generalisation performance for Digit /FIVE/ was remarkably improved, as well as Digit /NINE/.

It should be noted that, interestingly, the generalisation performance for the class(es) other than those with the attentive states was also improved (i.e. Digit /FIVE/ in Table 10.5 and Digit /THREE/ in Table 10.6). This may be considered as the "side-effect" of having the attentive states; since the pattern space for the digits with the attentive states was more consolidated, the coverage of the space for other digits accordingly became more accurate.

From these observations, it is considered that, since the performance improvement for Digit /NINE/ in both the cases was not more than expected, the pattern space for Digit /NINE/ is much harder to cover fully than other digits.

For both the OptDigit and PenDigit data sets, a similar performance improvement to the SFS case was obtained; for the OptDigit, the performance of Digit /NINE/ was relatively poor (57.5%), then the number of the RBFs within each of LTM Nets (2 to 3) for Digit /NINE/ was increased from 2 to 8 (which yields the total number of RBFs in LTM Nets 1 to 3, 77), and the performance for Digit /NINE/ was remarkably increased at 67.5%, which resulted in the overall generalisation performance of 87.5% (initially 87.0%).

Similarly, for the PenDigit, a performance improvement of 5.0% (i.e. from 80.0% to 85.0%) for Digit /NINE/ was obtained by increasing the number of RBFs from 4 to 6 in each LTM Net (2 to 5) for Digit /NINE/ only (then, the total number of RBFs in LTM Nets (1 to 5) is 183), which yielded the overall generalisation performance of 89.8% (i.e. initially 89.3%).

10.7 An Extension to the HA-GRNN Model – Implemented with Both the Emotion and Procedural Memory within the Implicit LTM Modules

In the previous section, it has been described that the model of HA-GRNN, which takes into account the concept of the four modules within the AMS, i.e. attention, intuition, LTM, and STM, can be applied to the intelligent pattern recognition system and thereby successfully contributed to a performance improvement in the pattern recognition context.

In this section, we consider another model (cf. Hoya, 2003d), which can be regarded as an extension to the HA-GRNN model.

Fig. 10.8 shows the architecture of the extended model. As in the figure, the two modules within the AMS context, i.e. the emotion and procedural part of implicit LTM (i.e. indicated by "Procedural Memory" in the figure), are also considered within the extended model, in comparison with the original HA-GRNN. It is considered that the ratio between the numbers of attentive and non-attentive kernel units within the STM is determined by the control mechanism, one part of which can be represented as (the functionality of) the **attention** module (see Sect. 10.2), and that, within the control mechanism, the perceptual output $\mathbf{y}$ is also temporarily stored. (Therefore, the control mechanism can also be regarded as a part of the **STM/working memory** or the associated module, such as **intention** or **thinking** (cf. Fig. 5.1 and see Sects. 8.3, 9.3, and 10.4). In addition, in Fig. 10.8, both the actuators and emotional expression mechanism can be dealt within the context of the **primary output** module of the AMS.)

In the figure, the input matrix $\mathbf{X}_{in} = [\mathbf{x}^1, \mathbf{x}^2, \ldots, \mathbf{x}^{N_s}]$ $(N_L \times N_s)$ is given as a collection of the sensory input vectors[10], where $\mathbf{x}^i = [x_1^i, x_2^i, \ldots, x_{N_L}^i]^T$ $(i = 1, 2, \ldots, N_s$, N_s: number of the sensory inputs) with length $N_L = \max(N_i)$. (Thus, for each column in $\mathbf{X}_{in}$, if $N_i < N_L$ a zero-padding operation is, for instance, performed to fill fully in the column.) Note that, since the STM, as well as the LTM (i.e. "Kernel Memory" (1 to L) and the procedural memory in Fig. 10.8) is based upon the kernel memory concept, it can simultane-

---

[10]Here, it is assumed that the input data are already acquired after the necessary pre-processing steps, i.e. via the cascade of pre-processing units in the **sensation** module within the AMS context (See Chap 6).

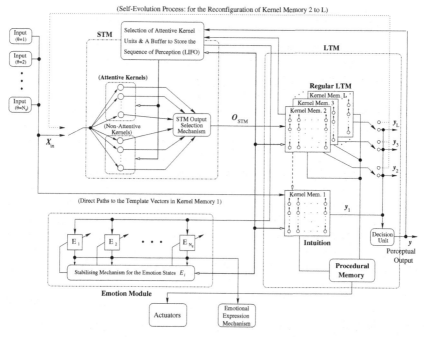

Fig. 10.8. An extension to the HA-GRNN model, with both the modules representing emotion (i.e. equipped with N_e emotion states) and procedural memory within the implicit LTM (i.e. indicated by "Procedural Memory"). Note that "Kernel Memory" (1 to L) within the extended model correspond respectively to LTM Nets (1 to L) within the original HA-GRNN (cf. Fig. 10.4); each kernel memory can be formed based upon the kernel memory principle (in Chaps. 3 and 4) and thus shares more flexible properties than PNNs/GRNNs. (Moreover, in the figure, two different types of the arrows are used; the arrows filled in black depict the actual data flows, whereas the ones filled with white indicate the control flows)

ously receive and then process the multi-modal input data $\mathbf{X}_{in}$ and eventually yields the STM output matrix $\mathbf{O}_{STM}$ $(N_L \times N_s)$ via the STM output selection mechanism. Then, the STM output matrix $\mathbf{O}_{STM}$ is presented to the LTM, resulting in the generation of the output vectors $\mathbf{y}_j = [y_j^1, y_j^2, \ldots, y_j^{N_s}]^T$ $(j = 1, 2, \ldots, L)$ from the respective kernel memory (1 to L). Eventually, similar to the HA-GRNN (cf. Fig. 10.4), the final output $\mathbf{y} = [y^1, y^2, \ldots, y^{N_s}]$ can be obtained from the decision unit (e.g. by following the "winner-takes-all" scheme) as the *perceptual* output (i.e. corresponding to the **secondary output** within the AMS context).

10.7.1 The STM and LTM Parts

As aforementioned, both the STM and LTM parts can be constructed based upon the kernel memory concept within the extended model;

- **STM**

  Is represented by a collection of kernel units and (partially)[11] the associated control mechanism. The kernel units within the STM are divided into the attentive and non-attentive kernels by the control mechanism.

- **LTM: Kernel Memory (2 to L)**

  Is considered as regular LTM. In practice, it is considered that each Kernel Memory (2 to L) is partitioned according to the domain/modality specific data. For instance, provided that the kernel units within Kernel Memory i ($i = 2, 3, \ldots, L$) are arranged in a matrix as in Fig. 10.9 (on the left hand side), the matrix can be sub-divided into several data-/modality-dependent areas (or submatrices).

- **LTM: Kernel Memory 1 (for Generating the Intuitive Outputs)**

  Is essentially the same as Kernel Memory (2 to L), except that the kernel units have the direct paths to the input matrix $\mathbf{X}_{in}$ and thereby can yield the intuitive outputs.

In both the STM and LTM parts, the kernel unit representation in Fig. 3.1, 3.2, or 10.3 is alternatively exploited. Then, in Fig. 10.9, provided that the kernel units within Kernel Memory i ($i = 1, 2, \ldots, L$) are arranged in a matrix as in Fig. 10.9 (on the left hand side)[12], the matrix can be sub-divided into several data-dependent areas (or sub-matrices). In the figure, each modality specific area (i.e. auditory, visual, etc) is represented by a column (i.e. the total number of columns can be equivalent to N_s; the total number of sensory inputs), and each column/sub-matrix is further sub-divided and responsible for the corresponding data sub-area, i.e. alphabetic/digit character or voice recognition (sub-)area, and so forth. (Thus, this somewhat simulates the PRS within the implicit LTM.)

Then, a total of N_s pattern recognition results can be obtained at a time from the respective areas of the i-th Kernel Memory (and eventually given as a vector $\mathbf{y}_i$).

Since the formation of both the STM and LTM parts can be followed by essentially the same evolution schedule as that of the HA-GRNN (i.e. from Phase 1 to Phase 4; see Sect. 10.6.3), it is expected that from the kernel units

---

[11] Compared with Fig. 5.1 (on page 84), it is seen that the control mechanism within the extended model (in Fig. 10.8) somewhat shares both the aspects of the two distinct modules, the **STM/working memory** (i.e. in terms of the temporal storage of the perceptual output) and **attention** module (i.e. for determining the ratio between the attentive and non-attentive kernel units; cf. the attended kernels in Sect. 10.2.1), within the AMS context. Thus, it is said that the associated control mechanism is partially related to the STM/working memory module.

[12] As described in Chap. 3, there is no restriction in the structure of kernel memory. However, here a matrix representation of the kernel units is considered for convenience.

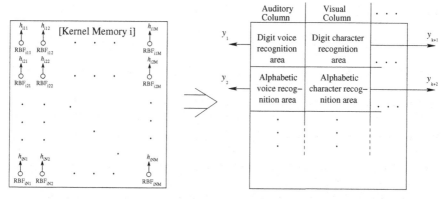

Fig. 10.9. The i-th Kernel Memory (in Fig. 10.8) arranged in a matrix form (left) and its division into several data-dependent areas/sub-matrices (right). In the figure, each modality specific area (i.e. for the auditory, visual, etc) is represented by a column (i.e. the total number of columns can be equivalent to N_s; the total number of sensory inputs), and each column/sub-matrix is further sub-divided and responsible for the corresponding data sub-area, i.e. alphabetic/digit character or voice recognition area, and so forth

within Kernel Memory 1^{13}, the pattern recognition results (i.e. provided that the model is applied to pattern recognition tasks) can be generated faster and more accurately (as observed in the simulation example of the HA-GRNN in Sect. 10.6.7).

Moreover, since these memory parts are constructed based upon the kernel memory concept, it is possible to consider that the kernel units are allowed to have not only the inter-layer (e.g. between the kernel units in Kernel Memory 2 and 3) but also cross-modality (or cross-domain) connections via the interconnecting link weights. Then, this can lead to more sophisticated data processing, e.g. simulating the mental imagery, where the activation(s) from some kernel units in one modality can occur without the input data but due to the transfer of the activation(s) from those in other modalities (e.g. the imagery of an object, via the auditory data → the visual counterpart; see also the simulation example of the simultaneous dual-domain pattern classification tasks using the SOKM in Sect. 4.5).

For the STM part, the procedure similar to that in the original HA-GRNN model (see Sect. 10.6.6), or alternatively, the general strategy of the **attention** module within the AMS (described in Sect. 10.2), can be considered for determining the attentive/non-attentive kernel units. In addition, the perceptual output **y** can be temporarily held within the associated control mechanism for both the attentive and emotion states to affect the determination.

---

[13] As described in the HA-GRNN, Kernel Memory 1 (i.e. corresponding to LTM Net 1) may be merely treated as a collection of the kernel units, instead of a distinct LTM module/agent, within the LTM part in the actual implementation. For this issue, see also Sect. 10.5.

10.7.2 The Procedural Memory Part

As discussed in Sect. 8.4.2, it is considered that some of the kernel units within Kernel Memory (1 to L) may also have established the connections (via the interconnecting link weights) with those in the procedural memory; due to the activation(s) from such kernel units, the kernel units within the procedural memory can be subsequently activated (via the link weights). Albeit dependent upon the manner of implementation, it is considered that each kernel unit within the procedural memory holds a set of control data which can eventually cause the corresponding motoric/kinetic actions from the body (i.e. indicated by the mono-directional link between the procedural memory and actuators in Fig. 10.8).

Then, the kernel units corresponding to the respective sets of control data (i.e. represented as a form of the template vector/matrix, e.g. to cause a series of the motoric/kinetic actions) can be pre-determined and installed within the procedural memory. In such a case, e.g. a chain of ordinary symbolic nodes may be sufficiently exploited. However, it is alternatively possible that such a sequence can be acquired via the learning process between the STM and LTM parts (i.e. represented by a chain of kernel units/kernel network(s); see also Chap. 7 and Sect. 8.3.2) and later transformed into the procedural memory (i.e. by exploiting the symbolic kernel unit representation in (3.11)):

[Formation of Procedural Memory]

Provided that a particular sequence of the motoric/kinetic actions is still not represented by the corresponding chain of (symbolic) nodes within the procedural memory, once the learning process is completed, the kernel network (or chain of kernel units) composed by (regular) kernel units is converted into a fixed network (or chain) using the symbolic node representation in (3.11). In practice, this can be helpful for saving the computation time in the data processing. However, when the kernel units are transformed into the corresponding symbolic nodes, the data held within the template vectors will be lost and therefore no longer accessible from the STM part.

Thus, within the extended model, the procedural memory can be viewed (albeit not limited to) as a collection of the chains of symbolic nodes so obtained.

10.7.3 The Emotion Module and Attentive Kernel Units

As in Fig. 10.8, the emotion module with 1) the emotional states E_i ($i = 1, 2, \ldots, N_e$) and 2) a stabilising mechanism for the emotional states is also considered within the extended model.

Then, for determining the attentive/non-attentive kernel units within the STM of the extended model, the embedded emotion states E_i can be considered as the criteria; despite that the attentive states (represented by the RBFs) were manually determined as those within the previous HA-GRNN model (i.e. see the simulation example in Sect. 10.6.7), the attentive/non-attentive kernel units can be autonomously set, depending upon the application.

For instance, we may implement the following strategy:

[Selecting the Attentive Kernel Units &
Updating the Emotion States E_i]

Step 1)
  Search a kernel unit(s) within the regular LTM part (i.e. Kernel Memory 2 to L) attached with the emotional state variables e_i ($i = 1, 2, \ldots, N_e$, assuming that the kernel unit representation in Fig. 10.3 is exploited), the values of which are *similar* to the current values of E_i. Then, set the kernel unit(s) so found as the attentive kernel units (via the control mechanism) within the STM.

Step 2)
  Then, whenever the kernel unit(s) within the LTM (i.e. Kernel Memory 1 to L) is activated by i.e. the incoming data $\mathbf{X}_{in}$ or transfer of other kernel units via the link weights, the current emotion states (at time n) $E_i(n)$ ($i = 1, 2, \ldots, N_e$) are updated by recalling the emotional state variables attached:

$$E_i(n + 1) = E_i(n) + \sum_{j=1}^{N_K} e_i^j(n) K_j \qquad (10.6)$$

  where N_K is the number of kernel units so activated, e_i^j correspond to the emotional state variables attached to such a kernel unit, and K_j is the activation level of the kernel unit.

Step 3)
  Continue the search for the kernel unit(s) in order to make E_i close to the optimal E_i^{*}[14], i.e.

$$\sum_{i=1}^{N_e} |E_i - E_i^{*}| \le \theta_E \qquad (10.7)$$

  where θ_E is a certain constant.

---

[14]In this strategy, only a single set of the optimal states E_i^{*} is considered, without loss of generality. These optimal states can then be regarded as the pre-set values defined in the innate structure module within the AMS context.

As in Step 1), the functionality of the control mechanism for the STM in Fig. 10.8 is to set the attentive and non-attentive kernel units, whilst it is considered that the stabilising mechanism for the emotion states plays the role for both Steps 2) and 3). (In Fig. 10.8, the latter is indicated by the signal flows between the stabilising mechanism and Kernel Memory 1 to L; see also Sect. 10.3.7.)

For the representation of the emotion states, the two intensity scales given in (10.1) and (10.2) can, for instance, be exploited for both E_1 and E_2 (or e_1 and e_2, albeit not limited to this representation). Then, the rest may be used for representing the current internal states of the body, imitating issues such as boredom, hunger, thirst, etc., depending upon the application. (Accordingly, the number of the emotional state variables attached to each kernel unit within the memory parts may be limited to 2.)

The optimal states E_i^* must be carefully chosen in advance dependent upon the application to achieve the goal; within the AMS context, this is relevant to the design of the **instinct: innate structure** module. In practice, however, it seems rather hard to consider the case where the relation (10.7) is satisfied, since, when it is active, i) the surrounding environment never stays still, thereby ii) the external stimuli (i.e. given as the input data $\mathbf{X}_{in}$ within the extended model) always affect the current emotion states E_i to a certain extent, and thus iii) (if any) the relation (10.7) does not hold that long.

Therefore, it is considered that the process for the selection of the attentive kernel units and updating the emotion states E_i will be continued endlessly, whilst it is active.

10.7.4 Learning Strategy of the Emotional State Variables

For the emotional state variables e_i attached to each kernel unit, the values may be either i) determined (initially) *a priori* or ii) acquired/varied via the learning process, depending upon the implementation.

For i), it is considered that the assignment of the variables may be necessary prior to the utility of the extended model; i.e. as indicated by the relationship (or the parallel functionality) between the **emotion** and **instinct: innate structure** module in Fig. 5.1, some of the emotional state variables must be pre-set according to the design of the instinct: innate structure module, whilst others may be dynamically varied, within the AMS context. (For some applications, this may involve rather laborious tasks by humans; as discussed in Sect. 8.4.6.)

In contrast, for ii), it is possible to consider that, as described earlier in terms of the implicit/explicit emotional learning (i.e. in Sects. 10.3.4 and 10.3.5, respectively), although the emotional state variables are initially set to the neutral states, the variables may be updated by the following strategy:

[Updating the Emotional State Variables]

For all the activated kernel units, update the emotional state variables e_i^j ($i = 1, 2, \ldots, N_e$):

$$e_i^j \leftarrow (1 - \lambda_e)e_i^j + \lambda_e E_i$$

$$\lambda_e = \lambda'_e \frac{E_i - E_{i,min}}{E_{i,max} - E_{i,min}} \qquad (10.8)$$

where $0 < \lambda'_e \leq 1$, E_i are the current emotion states of the extended model, and $E_{i,max}$ and $E_{i,min}$ correspond respectively to the maximum and minimum value of the emotion state.

Then, as described in terms of the evolutionary process of the HA-GRNN model (i.e. such as the STM $\longleftrightarrow$ LTM learning process; see Sect. 10.6.3), such activated kernel units may be eventually transferred/transformed into the LTM, depending upon the situation. (In particular situations, this can thus be related to the implicit/explicit emotional learning process as discussed in Sects. 10.3.4 and 10.3.5, respectively).)

In the late 1990's, an autonomous quadruped robot (named as "MUTANT") was developed (Fujita and Fukumura, 1996), in which the movement is controlled by a holistic model somewhat similar to the AMS, equipped with two sensory data (i.e. both the sound and image data, as well as the processing mechanism of the perceptual data) and the respective modules imitating such psychological functions as attention, emotion, and instinct. Subsequently, the emotionally grounded (EGO) architecture (Takagi et al., 2001), in which the two-stage memory system of STM and LTM is considered together with the aforementioned three psychologically-oriented modules, was developed for controlling the behaviour of the humanoid SDR-3X model (see also Ishida et al., 2001)/ethological robot of AIBO for entertainment (see also Fujita, 1999, 2000; Arkin et al., 2001), which led to a great success in that the robots were developed by fully exploiting the available (albeit rather limited range of) technologies and were generally accepted in world wide.

For each EGO or MUTANT, although the architecture is not shown fully in detail in the literature, it seems that both the models are rather based upon a conventional symbolic processing system and hence considered to be rather hard to develop/extend to more dynamic systems; in the MUTANT (Fujita and Fukumura, 1996), the module "automata" can be compared to the STM/working memory module (and/or the associated modules such as intention and thinking) of the AMS. However, it seems that, unlike the the AMS, the target behaviour of the robot is to a large extent pre-determined (i.e. not varied by the learning) based only upon the resultant symbol(s) obtained by applying the well-known Dijkstra's algorithm (Dijkstra, 1959), which globally finds the shortest path on a fixed graph (see e.g. Christofides,

1975) and is thus considered to be rather computationally expensive (especially when the number of nodes becomes larger). Therefore, it seems rather hard to acquire new patterns of behaviours through the learning process (since it seems that a static graph representation is used to bind a situation to a motion of the robot). Moreover, the attention mechanism also seems to be pre-determined; by the attention mechanism, the robot can only pay attention to a pre-determined set of sound or visual target and thereby move the head.

In contrast, although both the STM and LTM mechanisms are implemented within the EGO architecture, it seems that these memory mechanisms are not sufficiently plastic, since for the voice recognition, the HMM (see e.g. Rabiner and Juang, 1993; Juang and Furui, 2000) is employed, or it can suffer from various numerically-oriented problems, since such conventional ANNs as associative memory or HRNN (Hopfield, 1982; Hertz et al., 1991; Amit, 1989; Haykin, 1994) (see also Sect. 2.2.2) are considered within the mechanisms (Fujita and Takagi, 2003). Therefore, unlike the kernel memory, to adapt swiftly and at the same time robustly the memory system for time-varying situations is generally considered to be hard within these models.

10.8 Chapter Summary

In this chapter, we have focused upon the remaining four modules related to the abstract notions of mind, i.e. attention, emotion, intention, and intuition module, within the AMS.

Within the AMS context, the functionality of the four modules is summarised as follows:

- **Attention Module**:
  As described in Sect. 10.2.1, the attention module acts as a *filter* and/or *buffer* that picks out a particular set of data and holds temporarily the information about e.g. the activation pattern of some of the kernel units within the memory modules (i.e. the **STM working memory** or **LTM and/or oriented** modules), in order for the AMS to initiate a further memory search process (at an appropriate time, i.e. by the **thinking** or **intention** module) from the attended kernel units; in other words, a priority will be given for the memory search process amongst the marked kernel units by the STM/working memory module.
- **Emotion Module**:
  As described in Sect. 10.3.1, the emotion module has two aspects: the aspect of i) representing the current internal states of the body by a total of N_e emotion states within it, due to the relation with the **instinct: innate structure** and **primary output** modules, and that of ii) memory, i.e. as in Fig. 10.2 (or the alternative

kernel unit representation in Fig. 10.3), the kernel units within the STM/working memory/LTM modules are connected with the emotion module.

- **Intention Module**:
  Within the AMS, the intention module can be used to hold temporarily the information about the resultant states so reached during performing the thinking process by the **thinking** module. In reverse, the state(s) within the module can affect the manner of the thinking process to a certain extent. Although it may be seen that the functionality can be similar to that of the **attention** module, the duration of holding the state(s) is relatively longer and less sensitive to the incoming data arrived at the **STM/working memory** module than that within the attention module.
- **Intuition Module**:
  As described in Sect. 10.5, the intuition module can be considered as another **implicit LTM** module within the AMS, formed based upon a collection of the kernel units that have exhibited repetitively and relatively strong activations within the LTM/LTM-oriented modules. However, unlike the regular implicit LTM module, the activations from such kernel units may affect directly the thinking process performed by the **thinking** module.

Then, in the subsequent Sects. 10.6 and 10.7, the five modules within the AMS, i.e. attention, emotion, intuition, (implicit) LTM, and STM/working memory module, have been modelled and applied to develop an intelligent pattern recognition system. Through the simulation examples of HA-GRNN, it has then been observed that the recognition performance can be improved by implementing these modules.

11

Epilogue – Towards Developing A Realistic Sense of Artificial Intelligence

11.1 Perspective

So far, we have considered how the artificial mind system based upon the holistic model as depicted in Fig. 5.1 (on page 84) works in terms of the associated modules and their interactive data processing. It has then been described that most of the modules and the data processing can be represented in terms of the kernel memory concept. In the closing chapter, a summary of the modules and their mutual relationships is firstly given. Then, we take into account the enigmatic and (probably) the most controversial topic of consciousness within the AMS principle. Finally, we close the book by making a short note on the brain mechanism for intelligent robotics.

11.2 Summary of the Modules and Their Mutual Relationships within the AMS

In Chaps. 6–10, we considered in detail i) the respective roles of the 14 modules within the AMS, ii) how these modules are inter-related to each other, and iii) how they are represented by means of the kernel memory principle to perform the data processing, the principle of which has been described extensively in the first part of the book (i.e. in Chaps. 3 and 4).

In Chap. 5, it was described that the holistic model of the AMS (as illustrated in Fig. 5.1) can be macroscopically viewed as an input-output system consisting of i) one single input (i.e. the **sensation** module), ii) two output (i.e. the **primary output** and **secondary: perceptual output** modules), and iii) the other 11 modules, each representing the corresponding cognitive/psychological function.

Then, the functionality of the 14 modules within the AMS can be summarised as follows:

1) **Input: Sensation Module (Sect. 6.2)**
   Functions as the input mechanism for the AMS. It receives the sensory

Tetsuya Hoya: *Artificial Mind System – Kernel Memory Approach*, Studies in Computational Intelligence (SCI) **1**, 237–244 (2005)
www.springerlink.com

data from the outside world, converts them into the data which can be efficiently handled within the AMS, and then sends them to the STM/working memory module.

2) **Attention Module (Sect. 10.2)**

   Acts as a filter and/or a buffer which picks out a particular set of data and holds temporarily the information about the activated kernel units within the memory-oriented modules (i.e. explicit/implicit LTM, intuition, STM/working memory, and semantic networks/lexicon modules). Such kernel units are then regarded as attended kernel units and give priority to initiate a further memory search (at an appropriate period of time) via the intention/thinking module.

3) **Emotion Module (Sect. 10.3)**

   Inherently exhibits the two aspects, i.e. i) to represent the current (subset of) internal states of the body (due to the relationship with the instinct: innate structure/primary output module) and ii) memory in terms of the connections with the kernel units within the memory modules (or alternatively represented by the emotional state variables attached to them as shown in Fig. 10.3, on page 197).

4) **Explicit (Declarative) LTM Module (Sect. 8.4.3)**

   Is the part of the LTM, the contents of which can be accessible from the STM/working memory module, where required (i.e. the data flow **explicit LTM $\longrightarrow$ STM/working memory** in Fig. 5.1; hence the term *declarative*). The concept of the module is closely tied to that of the semantic networks/lexicon module. Within the kernel memory principle, it consists of multiple kernel units.

5) **Implicit (Nondeclarative) LTM Module (Sect. 8.4.2)**

   Is the part of the LTM which may represent the procedural memory, PRS, or non-associative learning (i.e. habituation and sensitisation). Unlike the explicit LTM, the contents within the module cannot be accessible from the STM/working memory module (hence the term *nondeclarative*). Within the kernel memory principle, it can be represented by multiple kernel units with directional data flows (i.e. for the mono-directional flow **STM/working memory $\longrightarrow$ implicit LTM**; see also Sect. 3.3.4).

6) **Instinct: Innate Structure Module (Sect. 8.4.6)**

   Can be regarded as a (rather static) part of the LTM; it may be composed by a collection of pre-set values (i.e. also represented by kernel units) which reflect e.g. the physical limitations/properties of the body and can be exploited for giving the target responses/reinforcement signals during the learning process of the AMS. Then, the behaviour of the AMS can be significantly affected by virtue of the module. In this respect, the instinct: innate structure module should be carefully taken into account for the design of the other associated modules such as emotion, input: sensation, implicit LTM, intuition, and language module.

7) **Intention Module (Sect. 10.4)**

The functionality of the module can be seen essentially similar to the attention module; the module can be used to hold temporarily the information about the resultant states reached by the thinking module, i.e. represented in terms of the activation pattern(s) of the kernel units within the memory-oriented modules. However, unlike the attention module, the state(s) within the intention module can in reverse affect the manner of the thinking process to a certain extent. Moreover, the duration of holding such state(s) is relatively longer and less sensitive to the incoming data which arrives at the STM/working memory module than the attention module.

8) **Intuition Module (Sect. 10.5)**

Can be considered as another implicit LTM (as described in Sect. 10.5) within the AMS. In terms of the kernel memory principle, it is formed based upon a collection of the kernel units that have exhibited repetitively and relatively strong activations within the LTM/LTM-oriented modules during the learning. However, unlike the regular implicit LTM, the activations from such kernel units can affect directly the manner of the data processing within the thinking module.

9) **Language Module (Sect. 9.2)**

Functions as a vehicle for the thinking process performed by the thinking module. The module can be defined as a built-in but dynamically reconfigurable learning mechanism, consisting of a set of grammatical rules represented in terms of the kernel memory principle. Hence, the module has a close relationship with the semantic networks/lexicon module.

10) **Semantic Networks/Lexicon Module (Sects. 8.4.4 and 9.2)**

Is considered as the semantic part of the (explicit) LTM (and hence is closely related to the explicit LTM and language modules, albeit depending upon the manner of implementation) within the AMS and, as other LTM-oriented modules, can be represented by the kernel memory.

11) **STM/Working Memory Module (Sect. 8.3)**

Plays the central part for performing various interactive data processing between other associated modules within the AMS. For instance, the incoming data received from the input: sensation module are temporarily held, converted to the respective kernel units, and may be eventually transformed into the kernel units within the LTM/LTM-oriented modules through the learning process (in Chap. 7). The kernel units within the STM/working memory module are also used for a further memory search/thinking process performed via the intention/thinking module.

12) **Thinking Module (Sect. 9.3)**

The module is considered to function in parallel with the STM/working memory and as a mechanism to organise the data processing (i.e. the memory search process within the memory-oriented modules) with the three associated modules, i.e. i) intention, ii) intuition, and iii) semantic networks/lexicon module. As described, one of such data processing performed via the thinking module is to determine the correctness of the sentence (e.g. represented by a chain of kernel units within the kernel memory context) in the semantical sense with the aid of the language module.

13) **Perceptual (Secondary) Output Module (Sect. 6.3)**
In the AMS context, the perception module is simply regarded as the output module that yields the secondary output of the AMS, which also represents the intermediate data processing occurring within the AMS, and the pattern recognition results obtained by accessing the contents of the LTM/LTM-oriented modules (such as the implicit LTM/intuition module) within the AMS. Thereby, such outputs are treated rather differently from the primary outputs within the AMS context.

14) **Primary Output Module (Sects. 9.3.3 and 10.3)**
Is the module directly connected to the physical devices for causing real actions, such as motions from the body or the internal activities i.e. simulating the endocrine system. Similar to the secondary (perceptual) outputs, the state(s) within the primary output module may be fed back to the STM/working memory module.

As in the above, the kernel memory concept, which was described extensively in Chaps. 3 and 4, plays a fundamental role to embody all the 14 modules within the AMS.

11.3 A Consideration into the Issues Relevant to Consciousness

To describe what is *consciousness* has historically been a matter of debate (see e.g. Turing, 1950; Terasawa, 1984; Dennett, 1988; Searle, 1992; Greenfield, 1995; Aleksander, 1996; Chalmers, 1996; Osaka, 1997; Pinker, 1997; Hobson, 1999; Shimojo, 1999; Gazzaniga et al., 2002). Although we all can inherently have the conscious experience, it is hard to define it. It has long been considered that consciousness is the key concept of so-called "mind-brain" research (or alternatively called the *ontological* problem within the philosophical context; see e.g. Gazzaniga et al. (2002)). There is, however, still no satisfactory understanding of consciousness.

In a cognitive scientific view, Pinker suggested a framework for thinking about the problem of consciousness (Pinker, 1997). In his theory, the problem of consciousness can be separated into the following three issues (Pinker, 1997; Gazzaniga et al., 2002):

- **Sentience**:
  This notion refers to subjective experience, phenomenal awareness, raw feelings, first person tenses, what it is like to be or do something. If you have to ask, you will never know.
- **Access to information**:
  The ability to report on the content of mental experience without the capacity to report on how the content was built up by the nervous system.
- **Self-knowledge**:
  Amongst the people and objects that an intelligent being can have accurate information about is the being itself.

Then, according to Gazzaniga et al. (Gazzaniga et al., 2002), the latter two may be dealt within the cognitive neuroscientific context. However, for the remaining one, they unanimously share a view that science has little to say about sentience. Moreover, Searle, a philosopher of our age, also claimed that the science will never understand the nature of subjective experience (Searle, 1992; Gazzaniga et al., 2002). The first is thus closely relevant to the encompassed issue of the so called *qualia*: why a physical system with a particular architecture gives rise to such *feelings* and (thus the term) *qualia* (Chalmers, 1996; Wilson and Keil, 1999).

On the other hand, despite these intangible issues, Turing denies that the question of consciousness has much relevance to the practice of AI, though he admits that the question of consciousness is a difficult one (Turing, 1950; Russell and Norvig, 2003): *"I do not wish to give the impression that I think there is no mystery about consciousness ... But I do not think these mysteries necessarily need to be solved before we can answer the question with which we are concerned in this paper."* In this regard, the philosopher of our age Dennett is also supportive (see the interview on pp. 658-659 in Gazzaniga et al., 2002).

In Chap. 5, albeit putting aside the rigorous justification of the above three issues pertinent to consciousness, it was proposed that the AMS consists of a total of 14 modules, each of which roughly corresponds to the element for describing the consciousness due to Hobson (Hobson, 1999), and the classification of the modules within the AMS into those functioning either with or without consciousness was made in a rather narrow sense; the modules classified as those functioning with consciousness, i.e. the five modules: attention, emotion (partially), intention, STM/working memory, and thinking module, may then be considered to correspond to the latter two issues, i.e. the issues of access to information and self-knowledge, whilst the other functioning without consciousness do not.

Then, it seems more appropriate that we will resume the discussion of consciousness, upon the embodiment and actual implementation of all the 14 modules to construct the entire AMS, which has yet to be done.

For this purpose, we may resort to the ordinary thought experiment; suppose that we have built an intelligent being which embodies the whole AMS (i.e. based upon either hardware or software, or even wetware) and, at the same time, developed an external device that can trace all data processing occurring within the AMS and output the results that can be perceived by us (e.g. visibly, at our desired level of data processing); we can not only obtain the results of e.g. the data processing from the external device but also change the observation level at the module (i.e. macroscopic) or kernel unit (i.e. microscopic) level, etc., at any moment. Also, suppose that the intelligent being is also equipped with a communication device (i.e. the voice generation mechanism) and thereby can communicate with us and report how it *feels* or *thinks*.

Then, we may be able to handle sufficiently with the two aforementioned issues ii) and iii), by matching the results obtained from the externally-located tracing device (i.e. the objective measurement) and the reports simultaneously obtained from the intelligent being by way of the communication between the intelligent being and ourselves (in this regard, the latter can be considered as the subjective measurement).

Nevertheless, for the time being, we may well follow the principle of Turing; the ultimate goal of this book has been to provide a direction/insight towards developing the artificial system that can *simulate* the functionalities of mind and thereby is implemented in a more realistic sense of AI/robotics. Therefore, we stop digging into the discussion of consciousness here and leave it to a further study.

11.4 A Note on the Brain Mechanism for Intelligent Robots

Before closing the book, we make a brief note on the brain mechanism for intelligent robots within the AMS principle in this section.

As proposed in the first part of this monograph, the kernel memory concept provides the basis for developing the various modules within the AMS which have been extensively described in the second part. It is then considered that the kernel unit represents the most fundamental element to compose the mechanism for any higher-order functionalities of the AMS. In this sense, the principle generally agrees with the philosophical view due to Dreyfus (Dreyfus, 1972), since a kernel unit itself can represent a pattern or concept or perform the operation of the similarity between the input and the data stored, which can eventually lead to the development of the computer system different from the currently prevailing Von-Neumann type computers.

Generally, although the holistic model of AMS is versatile and can be exploited for any kind of robotics/AI applications, it is not considered to be appropriate for developing from scratch the AI/robots which can imitate any behaviours of highly intelligent creatures or humans in every detail, in terms of the AMS.

As described earlier (in Sect. 8.4.6), for developing such a highly intelligent being, possibly the hardest part will then be how to actually design (or initially define the set of parameters for) the instinct: innate module and many parts within the associated modules of the AMS, such as the emotion, intuition (or some part of explicit/implicit LTM), language (as well as semantic networks/lexicon and thinking), and (some part of) sensation module (albeit depending upon the applications).

This is partly because we still do not know exactly how to divide/specify the pre-determined part and the part that has to be self-evolved during the exposition to the outside world of the AMS and the associated learning process. Then, for another reason, this is since the amount/choice of such pre-set values (or "pre-wiring" process) by humans still may be prohibitively large or complicated, in order to reach such a high level of intelligence.

The former is hence somewhat relevant to the indication by Wiener (see Section IX: "On Learning and Self-Reproducing Machines" in Wiener, 1948): *"... we must ask for what we really want and not for what we think we want. The new and real agencies of the learning machine are also literal-minded ..."*, which implies that the designer (i.e. we) must know precisely in advance/predict how the AI/robot (is expected to) behaves in real situations, though it is thought that the learning capability inherently equipped within the AMS can greatly facilitate this and, eventually, even get rid of this dilemma. (Then, it somewhat reminds us of the considerable time, as well as the energy, spent so far for the learning then evolution of the real life, i.e. billions of years in order to adapt to the surrounding environment for the continuous life and preservation of the species, through countless numbers of iterations for the heuristic operations.)

On the other hand (and perhaps more crucially), although the above is only relevant to the issue of AMS, the technology currently available still is far from that for enabling us to embody the physical body of such autonomous robotics, which can act like a real creature in any detail with full controllability or the computing devices for modelling (conveniently) the respective functionalities of the AMS, as well as the interfaces to the body.

However, for limited purposes, it may be sufficient to exploit not all but only some of the modules, each based upon a much simpler architecture, as we have seen to develop the intelligent pattern recognition agents in Sects. 10.6 and 10.7.

Therefore, we at first should aim for the development of an intelligent agent that can simulate a limited set of the behaviours of the creatures within the AMS context, in parallel with the advancement in the technologies to embody the physical body (as well as the measurement technology).

In conclusion, as is usually the case for general engineering, one of the pragmatic ways may perhaps be to start developing a relatively small agent with a limited capacity by exploiting the currently available technology (i.e.

both the hardware and software) and thereby embodying a few of the modules within the AMS for a specific purpose (as in the intelligent pattern recognisers in the previous chapter), unleash it into the real environment, observe, then analyse its behaviour, and thereafter gradually make it more complex with adding other functionalities into the agent/robot, within the uniformed context.

References

Aleksander, I. (1996). *Impossible Minds: My Neurons and My Consciousness.* London: Imperial College Press.

Amari, S. (1967). Theory of adaptive pattern classifiers. *IEEE Trans. Electronic Computers*, EC-16, 299-307.

Amit, D. J. (1989). *Modeling Brain Function: The World of Attractor Neural Networks.* New York: Cambridge Univ. Press.

Anderson, A. K., Spencer, D. D., Fulbright, R. K., & Phelps, E. A. (2000). Contribution of the anteromedial temporal lobes to the evaluation of facial emotion. *Neuropsychology*, 14, 526-536.

Anderson, J., Platt, J. C., & Kirk, D. B. (1993). An analog VLSI chip for radial basis functions. in S. J. Hanson, J. D. Cowan, and C. L. Giles (Eds.) *Advances in Neural Information Processing Systems*, 5, 765-772.

Anderson, J. R. (1993). *Rules of the Mind.* Hillsdale, NJ:Eribaum.

Anderson, J. R. (2000). *Learning and Memory.* New York: John Wiley & Sons, Inc.

Apolinario, J. A., de Campos, M. L. R., & Diniz, P. S. R. (1997). Convergence analysis of the binormalized data-reusing LMS algorithm. *Proc. of the European Conf. Circuit Theory and Design*, Budapest, Hungary, 972-977.

Arkin, R. C., Fujita, M., Takagi, T., & Hasegawa, R. (2001). Ethological modeling and architecture for an entertainment robot. *Proc. of 2001 IEEE Int. Conf. Robotics & Automation*, 453-458, Seoul, Korea.

Arnold, M. B. & Gasson, J. (1954). Feelings and emotions as dynamic factors in personality integration. In M. B. Arnold and J. Gasson (Eds.). *The Human Person.* New York: Ronald, 294-313.

Asano, F., Hayamizu, S., Yamada, T., & Nakamura, S. (2000). Speech enhancement based on the subspace method. *IEEE Trans. Speech, Audio Processing*, 8-5, 497-507.

Asimov, I. (1950). *I, Robot.* New York: Doubleday, Garden City.

Atkinson, R. C. & Shiffrin, R. M. (1968). Human memory: a proposed system and its control processes. In K. W. Spence and J. T. Spence (Eds.), *The*

Psychology of Learning and Motivation, 2, 89-115, New York: Academic Press.

Badeau, R., Richard, G., & David, B. (2004). Sliding window adaptive SVD algorithms. *IEEE Trans. Signal Processing*, 52-1, 1-10.

Baddeley, A. D. & Hitch, G. (1974). Working memory. In G. H. Bower (Ed.), *The Psychology of Learning and Motivation*, 8, 47-89. New York: Academic Press.

Baddeley, A. D. (1986). *Working Memory*. Oxford: Oxford Univ. Press.

Baddeley, A., Gathercole, S., & Papagno, C. (1998). The phonological loop as a language learning device. *Psychological Review*, 105-1, 158-173.

Barrett, A. M., Crucian, G. P., Raymer, A. M., & Heilman, K. M. (1997). Spared comprehension of emotional prosody in a patient with global aphasia. *J. Int. Neuropsychol. Soc.*, 3, 57.

Barros, A. K., Kawahara, H., Cichocki, A., Kajita, S., Rutkowski, T., & Ohnishi, N. (2000). Enhancement of a speech signal embedded in noisy environment using two microphones. *Proc. of Int. Conf. Independent Component Analysis and Blind Signal Separation*, 423-428.

Barros, A. K., Rutkowski, T., Itakura, F., & Ohnishi, N. (2002). Estimation of speech embedded in a reverberant and noisy environment by independent component analysis and wavelets. *IEEE Trans. Neural Networks*, 13-4, 888-893.

Belouchrani, A., Abed-Meraim, K., Cardoso, J.-F., & Moulines, E. (1993). Second-order blind separation of temporally correlated sources. *Proc. of Int. Conf. Digital Signal Processing*, Cyprus, 346-351.

Bishop, C. M. (1996). *Neural Networks for Pattern Recognition*. Oxford: Oxford Univ. Press.

Borod, J. C., Koff, E., Perlman Lorch, M., & Nicholas, M. (1986). The expression and perception of facial emotion in brain damaged patients. *Neuropsychologia*, 24, 169-180.

Bowers, D., Bauer, R. M., & Heilman, K. (1993). The nonverbal affect lexicon: Theoretical perspectives from neuropsychological studies of affect perception. *Neuropsychology*, 7, 433-444.

Bowlby, J. (1971). *Attachment and Loss, Vol. 1: Attachment*. London: Hogarth.

Brian, S., Syrus, C. N., Rex, K., Raja, H., & Robert, H. N. (2001). A biologically motivated solution to the cocktail party problem. *Neural Computation*, 13-7, 1575-1602.

Broadbent, D. A. (1970). Stimulus set and response set: two kinds of selective attention. In D. I. Motofsky (Ed.). *Attention: Contemporary Theory and Analysis*. 51-60. New York: Appleton-Century-Crofts.

Brodal, A. (1982). *Neurological Anatomy*. New York: Oxford Univ. Press.

Brodmann, K. (1909). *Vergleichende Lokalisationslehre der Grosshirnrinde in ihren Prinzipien dargestellt auf Grund des Zellenbaues*. Leipzig: J. A. Barth. In G. von Bonin, *Some Papers on the Cerebral Cortex*. 201-230. Translated as, *On the Comparative Localization of the Cortex.* Springfield, IL: Charles C. Thomas, 1960.

Broomhead, D. S. & Lowe, D. (1988). Multivariable functional interpolation and adaptive networks. *Complex Systems*, 2, 321-355.

Bryson, A. E. & Ho, Y.-C. (1969). *Applied Optimal Control.* New York: Blaisdell.

Carpenter, G. A., Grossberg, S., & Reynolds, J. H. (1991). ARTMAP: Supervised real-time learning and classification of nonstationary data by a self-organizing neural network. *Neural Networks*, 4-5, 565-588.

Chalmers, D. (1996). *The Conscious Mind: In Search of a Fundamental Theory.* Oxford: Oxford Univ. Press.

Changeux, J. P. & Danchin, A. (1976). Selective stabilization of developing synapses as a mechanism for the specification of neural networks. *Nature*, 264, 705-712.

Chomsky, N. (1957). *Syntactic Structures.* Mouton.

Christianini, N. & Taylor, J. S. (2000). *An Introduction to Support Vector Machines and Other Kernel-Based Learning Methods*, Cambridge: Cambridge Univ. Press.

Christiansen, M. H. & Chater, N. (1999). Connectionist natural language processing: the state of the art. *Cognitive Sci.* 23-4, 417-437.

Christofides, N. (1975). *Graph Theory: An Algorithmic Approach.* Academic Press.

Cichocki, A., Gharieb, R. R., & Hoya, T. (2001). Efficient extraction of evoked potentials by combination of Wiener filtering and subspace methods. *Proc. of IEEE Int. Conf. Acoust. Speech, Signal Processing (ICASSP-2001)*, 5, 3117-3120.

Cichocki, A. & Amari, S. (2002) *Adaptive Blind Signal and Image Processing – Learning Algorithms and Applications*, John Wiley & Sons.

Colla, V., Sgarbi, M., Reyneri, L. M., & Sabatini, A. M. (1998). A neural approach to a sensor fusion problem. *Proc. of European Symp. Artificial Neural Networks (ESANN 1998)*, 357-362, Belgium.

Crane, T. (1995). *The Mechanical Mind: A Philosophical Introduction to Minds, Machines, and Mental Representation.* Penguin Books. Japanese translation: Tokyo: Keisou, Publishing, Co. Ltd.

Crochiere, R. E. & Rabiner, L. R. (1983). *Multirate Digital Signal Processing.* Englewood Cliffs, NJ: Prentice-Hall.

Darwin, C. (1872). *The Expressions of the Emotions in Man and Animals.* London: Murray.

Davis, G. M. (2002). *Noise Reduction in Speech Applications.* Florida: CRC Press.

Davidson, R. J., Ekman, P., Saron, C., Senulis, J., & Friesen, W. V. (1990). Approach/withdrawal and cerebral asymmetry: emotional expression and brain physiology. *J. Pers. Soc. Psychol.*, 38L, 330-341.

Dayhoff, J. E. & Gerstein, G. L. (1983). Favored patterns in nerve spike trains – I. Detection. *J. Neurophys..* 49 (6), 1334-1348.

Deller, Jr. J. R., Proakis, J. G., & Hansen, J. H. L. (1993). *Discrete-Time Processing of Speech Signals.* New York: Macmillan.

Dendrinos, M., Bakamidis, S., & Carayannis, G. (1991). Speech enhancement from noise: a regenerative approach. *Speech Communication*, 10, 45-57.

Dennett, D. C. (1984). Cognitive wheels: the frame problem of AI. In C. Hookway (Ed.), *Minds, Machines, and Evolution: Philosophical Studies.* 129-151, Cambridge: Cambridge Univ. Press.

Dennett, D. C. (1988). *Consciousness Explained.* Boston: Little Brown.

Descartes, R. (1984-5). *Philosophical Writings.* 3 vols. Trans. J. Cottingham, R. Stoothoff, and D. Murdoch. Cambridge: Cambridge Univ. Press.

Desimone, R., Albright, T. D., Gross, C. G., & Bruce, C. (1984). Stimulus-selective properties of inferior temporal neurons in the macaque. *J. Neurosci.*, 4, 2051-2062.

Dijkstra, E. W. (1959). A note on two problems in connection with graphs. *Numerische Mathematik*, 1, 269.

Ding, S., Hoya, T., Zhu, X., Barros, A. K., Daming, W., & Cichocki, A. (2004). Convolutive blind source separation of acoustic signals based on complex independent component analysis in the time-frequency domain and neural memory, in prepration for publication.

Doclo, S. & Moonen, M. (2000). Multi-microphone noise reduction using GSVD-based optimal filtering with ANC postprocessing stage. *Proc. of 9th IEEE Digital Signal Processing Workshop*, Hunt TX, USA.

Doclo, S. & Moonen, M. (2002). GSVD-based optimal filtering for single and multimicrophone speech enhancement. *IEEE Trans. Signal Processing*, 50-9, 2230-2244.

Douglas, S. & Cichocki, A. (1997). Neural networks for blind decorrelation of signals. *IEEE Trans. Signal Processing*, 45-11, 2829-2842.

Dreyfus, H. L. (1972). *What Computers Can't Do – the Limits of Artificial Intelligence.* Harper & Row, Publishers, Inc. Japanese Translation: *Computer-Niwa-Naniga-Dekinai-Ka?.* Sangyo-Tosho, Publishing, Co. Ltd.

Duda, R. O., Hart, P. E., & Stork, D. G. (2001). *Pattern Classification.* 2nd Ed., New York: Wiley.

Dudek, S. M. & Bear, M. F. (1992). Homosynaptic long-term depression in area CA1 of hippocampus and effects of N-methyl-D-aspartate receptor blockade. *Proc. of Natl. Acad. Sci. USA*, 89: 4363-4367.

Edelman, G. M. (1992). *Bright Air, Brilliant Fire.* Basic Books, Inc.

Ekman, P. (1971). Universals and cultural differences in facial expression. In J. K. Cole (Ed.), *Nebraska Symp. and Motivation*, 207-284. Licoln, NE: Univ. Nebraska Press.

Elman, J. L. (1990). Finding structure in time. *Cognitive Sci.* 14, 179-211.

Ephraim, Y. & Trees, H. L. V. (1995). A signal subspace approach for speech enhancement. *IEEE Trans. Speech, Audio Processing*, 3-4, 251-266.

Fodor, J. A. (1983). *The Modularity of Mind: An Essay on Faculty Psychology.* Cambridge: The MIT Press.

Forney, G. D. (1973). The Viterbi algorithm. *Proc. of IEEE*, 61, 268-278.

Forsyth, N., Chambers, J. A., & Naylor, P. A. (1999). A noise robust alternating fixed-point algorithm for stereophonic acoustic echo cancellation. *Electronics Letters*, 35-21, 1812-1813.

Freud, S. (1966). Project for a scientific psychology. Preliminary communication (to *Studies in Hysteria*, with (Josef Brauer). Three essays on sexuality. The unconscious, instincts, and their vicissitudes. The ego and the Id. All can be found in *The Complete Psychological Works of Sigmund Freud*. James Strachey. (Ed.24), vol. London: The Hogarth Press.

Fujita, M. & Fukumura, N. (1996). ROBOT entertainment. *Proc. of 6th Sony Research Forum*, 234-239 (in Japanese).

Fujita, M. (1999). Emotional expressions of a pet-type robot. *J. Robotics Soc. Japan*, 17-7, 947-951 (in Japanese).

Fujita, M. (2000). Digital creatures for future entertainment robotics. *Proc. of 2000 IEEE Int. Conf. Robotic & Automation*, 801-806, San Francisco, CA.

Fujita, M. & Takagi, T. (2003). Patent Application No. 2003-334785, Japan.

Furui, S. (1981). Cepstral analysis technique for automatic speaker verification. *IEEE Trans. Acoustic Speech and Signal Processing*, 29, 254-272.

Fukushima, K. (1975). Neocognitron: a self-organizing multilayered neural network. *Biological Cybernetics*, 20, 121-136.

Garcia, A. L. (1994). *Probability and Random Processes for Electrical Engineering*. 2nd Ed., Reading: Addison-Wesley.

Gazzaniga, M. S., Ivry, R. B., & Mangun, G. R. (2002). *Cognitive Neuroscience – the Biology of the Mind*, 2nd Ed., New York: W. W. Norton & Company.

Gold, B. & Morgan, N. (2000). *Speech and Audio Signal Processing*. John Wiley & Sons.

Golub, G. H. & Van Loan, C. F. (1996). *Matrix Computations*. 3rd Ed., Johns Hopkins Univ. Press.

Greenfield, S. A. (1995). *Journey to the Centers of Mind*, New York: W. H. Freeman and Company.

Gross, C. G., Rocha-Miranda, C. E., & Bender, D. B. (1972). Visual properties of neurons in inferotemporal cortex of the macaque. *Journal of Neurophysiology*, 35, 96-111.

Grossberg, S. (1988). *Neural Networks and Natural Intelligence*. Cambridge, MA: The MIT Press.

Gustafsson, H., Nordholm, S., & Claesson, I. (1999). Spectral subtraction using dual microphones. *Proc. of Int. Workshop on Acoustic Echo and Noise Control*, 60-63, Pennsylvania, U.S.A.

Gustafsson, H. *et al.*. (2003). System and method for dual microphone signal noise reduction using spectral subtraction. U.S.A. Patent 6549586, Apr. 2003.

Hand, D. J. (1984). *Kernel Discriminant Analysis*. Research Studies Press.

Hansen, P. C. & Jensen, S. H. (1998). FIR filter representation of reduced-rank noise reduction. *IEEE Trans. Signal Processing*, 46-6, 1737-1741.

Hansen, P. S. K. (1997). Signal subspace methods for speech enhancement. Ph.D. Thesis, Technical Univ. of Denmark, Lyngby, Denmark.

Hastie, T., Tibshirani, R., & Friedman, J. (2001). *The Elements of Statistical Learning*. New York: Springer-Verlag.

Haykin, S. (1994). *Neural Networks: A Comprehensive Foundation*. New York: Macmillan.

Haykin, S. (1996). *Adaptive Filter Theory*. Prentice-Hall, Inc.

Haykin, S. (2000). *Unsupervised Adaptive Filtering*. vol. I & II. John Wiley & Sons, Inc.

Hearst, M. A., Scholkopf, B., Dumais, S., Osuna, E., & J. Platt., J. (1998). Trends and controversies – support vector machines, *IEEE Intelligent Systems*, 13-4, 18-28.

Hebb, D. O. (1949). *Organization of Behavior*. New York: Wiley.

Hecht-Nielsen, R. (1998). A theory of the cerebral cortex. *Proc. of Int. Conf. Neural Info. Process. (ICONIP'98)*, 1459-1464. Burke, VA:IOS Press.

Heilman, K. M., Scholes, R., & Watson, R. T. (1975). Auditory affective agnosia: Disturbed comprehension of affective speech. *J. Neuro. Neurosurg. Psychiatry*, 38, 69-72.

Heinke, D. & Humphreys, G. W. (in press). Computational models of visual selective attention: A review. In G. Houghton (Ed.). *Connectionist Models in Psychology*, London: Psychology Press.

Heinke, D. & Humphreys, G. W. (in press). Attention, spatial representation and visual neglect: Simulating emergent attentional processes in the selective attention for identification model (SIAM). *Psychological Review*.

Hertz, J., Krogh, A., & Palmer, R. G. (1991). *Introduction to the Theory of Neural Computation*. Reading, MA: Addison-Wesley.

Hikosaka, O., Miyachi, S. Miyashita, K., & Rand, M. K. (1996). Procedural learning in monkeys – possible roles of the basal ganglia. In *Perception, Memory and Emotion: Frontiers in Neuroscience*, eds. T. Ono, B. L. McNaughton, S. Molotchnikoff, E. T. Rolls, and H. Nishijo, Elsevier, 403-420.

Hirsh-Pasek, K. & Golinkoff, R. M. (1996). *The Origins of Grammar: Evidence from Early Language Comprehension*. The MIT Press.

Hobson, J. A. (1999). *Ishiki-To-Nou (Consciousness and Brain)*. Tokyo: Tuttle-Mori Agency, Inc. & New York: W. H. Freeman and Company.

Hopfield, J. J. (1982). Neural networks and physical systems with emergent collective computational abilities. *Proc. of National Academy of Sciences of the U.S.A.* 81, 3088-3092.

Hoshino, O., Kashimori, Y., & Kambara, T. (1998). An olfactory recognition model based on spatio-temporal encoding of odor quality in the olfactory bulb. *Biological Cybernetics*, 79, 109-120.

Hovland, C. I. (1951). Human learning and retention. In S. S. Stevens (Ed.), *Handbook of Experimental Psychology*, 613-689, New York: John Wiley & Sons.

Hoya, T. (1998). Graph theoretic techniques for pruning data and their applications. *IEEE Trans. Signal Processing*, 46-9, 2574-2579.

Hoya, T. & Chambers, J. A. (2001a). Heuristic pattern correction scheme using adaptively trained generalized regression neural networks. *IEEE Trans. Neural Networks*, 12-1, 91-100.

Hoya, T. (2001b). Modeling the notions of intuition and consciousness by hierarchically arranged generalised regression neural networks. *Proc. of 2001 Int. Symp. Nonlinear Theory and Its Applications (NOLTA2001)*, 2, 403-406, Zao, Japan.

Hoya, T. (2003a). On the capability of accommodating new classes within probabilistic neural networks. *IEEE Trans. Neural Networks*, 14-2, 450-453.

Hoya, T., Cichocki, A., Tanaka, T., Hori, G., Murakami, T., & Chambers, J. A. (2003b). A combined cascading subspace methods and adaptive signal enhancement for stereophonic noise reduction. *Proc. of Fourth Int. Symp. Independent Component Analysis and Blind Signal Separation (ICA2003)*, 573-578, Nara, Japan.

Hoya, T., Barros, A. K., Rutkowski, T., & Cichocki, A. (2003c). Speech extraction based upon a combined subband independent component analysis and neural memory. *Proc. of Fourth Int. Symp. Independent Component Analysis and Blind Signal Separation (ICA2003)*, 355-360, Nara, Japan.

Hoya, T. (2003d). A kernel based neural memory concept and representation of procedural memory and emotion. *Proc. of 8th Int. Symp. Artificial Life and Robotics (AROB'03)*, 373-376, Oita, Japan.

Hoya, T. (2004a). Self-organising associative kernel memory for multi-domain pattern classification. *Proc. of IFAC Workshop on Adaptation and Learning in Control and Signal Processing (ALCOSP2004)*, 735-740, Yokohama, Japan.

Hoya, T. (2004b). Notions of intuition and attention modeled by a hierarchically arranged generalized regression neural network. *IEEE Trans. Systems, Man, and Cybernetics – Part B: Cybernetics*, 34-1, 200-209.

Hoya, T., Tanaka, T., Murakami, T., & Cichocki, A. (2004c). Stereophonic noise reduction by a combined multi-stage sliding subspace projection and adaptive signal enhancement. *Proc. of IFAC Workshop on Adaptation and Learning in Control and Signal Processing (ALCOSP2004)*, 421-426, Yokohama, Japan.

Hoya, T., Tanaka, T., Cichocki, A., Murakami, T., Hori, G., & Chambers, J. A. (2005). Stereophonic noise reduction using a combined sliding subspace projection and adaptive signal enhancement. *IEEE Trans. Speech and Audio Processing*, 13-3, 309-320.

Howells, P. W. (1976). Explorations in fixed and adaptive resolution at GE and SURC. *IEEE Trans. Antennas Propag.*, AP-24, Special Issue on Adaptive Antennas, 575-584.

Hubel, D. H. & Wiesel, T. N. (1977). The Ferrier lecture: functional architecture of macaque monkey visual cortex. *Proc. of R. Acad. Lond.*, Series B 198, 1-59.

Huckvale, M. (1996). *Speech Filing System Vs3.0 – Computer Tools For Speech Research*, London: University College.

Hudson, J. E. (1981). *Adaptive Array Principles.* Stevenage, U.K.: Peter Peregrinus.

Hugonnet, C. & Walder, P. (1998). *Stereophonic Sound Recording – Theory and Practice.* John Wiley & Sons.

Ishida, T., Kuroki, Y., Yamaguchi, J., Fujita, M., & Doi, T. (2001). Motion entertainment by a small humanoid robot based on OPEN-R. *Proc. of 2001 IEEE/RSJ Int. Conf. Intelligent Robots and Systems,* 1079-1086, Hawaii.

James, W. (1884). What is an emotion? *Mind,* 9, 188-205.

James, W. (1890). *The Principles of Psychology,* New York: Holt, Rinehart and Winston.

Jensen, S. H., Hansen, P. C., Hansen, S. D., & Sorensen, J. A. (1995). Reduction of broad-band noise in speech by truncated QSVD. *IEEE Trans. Speech, Audio Processing,* 3, 439-448, 1995.

Juang, B.-H. & Furui, S. (2000). Automatic recognition and understanding of spoken language – A first step toward natural human-machine communication. *Proc. of IEEE,* 88-8, 1142-1165.

Jutten, C. & Herault, J. (1991). Blind separation of sources, part I: an adaptive algorithm based on neuromimetic architecture. *Signal Processing,* 24-1, 1-10.

Jutten, C. (1997). Supervised composite networks. In E. Fiesler and R. Beale (Eds.), *Handbook of Neural Computation,* Chapter C1.6, New York: IOP Publishing and Oxford Univ. Press.

Karjalainen, P. A., Kaipio, J. P., Koistinen, A. S., & Vuhkonen, M. (1999). Subspace regularization method for the single-trial estimation of evoked potentials. *IEEE Trans. Biomed. Eng.,* 46-7, 849-860.

Kawato, M. (1996). *Nou-no Keisan Riron (Computational Theory of Brain).* Tokyo: Sangyo-Tosho, Co. Ltd.

Kawato, M., Doya, K., & Haruno, M. (2000). Gengo-Ni-Semaru-Tameno-Joken (Conditions towards language). *Kagaku,* 70, 381-387 (in Japanese).

Kenbo, H., Kindaichi, H., Shibata, T., Yamada, T., & Kindaichi, K. (Eds.) (1981). *The Japanese Dictionary: 3rd Edition.* Sanseido, Co. Ltd.

Kinoshita, J. (1996). *Neural-Network-to Gengo-Bunpoh (Neural Network and Language Grammar).* Japan:Kiku-chu, Publishing, Co. Ltd. (in Japanese)

Ko, C. C. & Siddharth, C. S. (1999). Rejection and tracking of an unknown broadband source in a two-element array through least square approximation of inter-element delay. *IEEE Signal Processing Let..* 6-5, 122-125.

Kitamura, T., Otsuka, Y., & Nakao, T. (1995). Imitation of animal behavior with use of a model of consciousness – behavior relation for a small robot. *Proc. of 4th. IEEE Int. Workshop on Robot and Human Communication,* 313-316, Tokyo.

Kitamura, T. (2000). *Robot-Wa-Kokoro-Wo-Motsuka? (Can Robots Have the Mind?).* Tokyo: Kyoritsu Publishing, Co. Ltd.

Kobayashi, T. & Kuriki, S. (1999). Principal component elimination method for the improvement of S/N in evoked neuromagnetic field measurements. *IEEE Trans. Biomed. Eng.* 46, 951-958.

Koch, C. (1999). *Biophysics of Computation*. Oxford Univ. Press.

Kohonen, T. (1997). *Self-Organizing Maps*. Berlin: Springer-Verlag.

Kolers, P. A. (1976). Reading a year later. *J. Exper. Psychol.: Human Learn. Memory*, 2, 554-565.

Kotter, R. & Meyer, N. (1992). The limbic system: A review of its empirical foundation. *Behav. Brain Res.*, 52, 105-127.

Kruschke, J. K. (1992). ALCOVE: An exemplar-based connectionist model of category learning. *Psychological Review*, 99-1, 22-44.

Lang, K. J. & Hinton, G. E. (1988). The development of the time-delay neural network. Technical Report CMU-CS-88-152, Carnegie-Mellon Univ.

R. Le Bouquin-Jennes, R., Akbari Azirani, A., & Faucon, G. (1997). Enhancement of speech degraded by coherent and incoherent noise using a cross-spectral estimator. *IEEE Trans. Speech, Audio Processing*, 5-5, 484-487.

Lee, C. H., Rabiner, L. R., Pieraccini, R., & Wilpon, J. G. (1990). Acoustic modeling for large vocabulary speech recognition. *Computer Speech and Language*, 4, 1237-65.

Le Doux, J. E. (1991). Emotion and the limbic system concept. *Concepts Neurosci.*, 2, 169-199.

Levelt, W. J. M. (1989). *Speaking: From Intention to Articulation*. Cambridge, MA: The MIT Press.

Looney, C. G. (1997). *Pattern Recognition Using Neural Networks – Theory and Algorithms for Engineers and Scientists*. New York: Oxford Univ. Press.

Low, R. & Togneri, R. (1998). Speech recognition using the probabilistic neural network. *Proc. of Int. Conf. Spoken Language Processing*, Paper No. 645, Sydney, Australia.

Lysetskiy, M., Lozowski, A., & Zurada, J. M. (2002). Invariant recognition of spatio-temporal patterns in the olfactory system model. *Neural Processing Letters*, 15, 225-234.

MacLean, P. D. (1949). Psychosomatic disease and the "visceral brain": Recent developments bearing on the Papez theory of emotion. *Psychosom. Med.*, 11,338-353.

MacLean, P. D. (1952). Some psychiatric implications of physiological studies on frontotemporal portion of limbic system (visceral brain). *Electroencephalogr. Clin. Neurophysiol.*, 4, 407-418.

MacQueen, J. B. (1967). Some methods for classification and analysis of multivariate observations. In *Proc. of Symp. Matho. Stat. Prob.*, 5th ed. Berkeley, CA:Univ. of Calif. Press, 1, 281-297.

Mak, M. W., Allen, W. G., & Sexton, G. G. (1994). Speaker identification using multilayer perceptron and radial basis function networks. *Neurocomputing*, 6, 99-117.

Mallat, S. (1999). *A Wavelet Tour of Signal Processing*. Academic Press.

Mandic, D. P. & Chambers, J. A. (2001). *Recurrent Neural Networks for Prediction: Learning Algorithms, Architectures, and Stability*. Chichester: John Wiley & Sons.

Martin, R. (1994). Spectral subtraction based on minimum statistics. *Proc. of EUSIPCO-94*, 1182-1185, Edinburgh.

Martin, R. (2001). Noise power spectral density estimation based on optimal smoothing and minimum statistics. *IEEE Trans. Speech, Audio Processing*, 9-5, 504-512.

Maruyama, K. (1981). *Saussure-No-Shiso (The Thought of Saussure)*. Tokyo: Iwanami Publishing, Co. Ltd. (in Japanese).

Matsumoto, G., Shigematsu, Y., & Ichikawa, M. (1995). The brain as a computer. In *Proc. of Int. Conf. Brain Processes, Theories & Models*, Cambridge, MA: The MIT Press.

McCarthy, J. & Hayes, P. J. (1969). Philosophical problems from the standpoint of artificial intelligence. *Machine Intelligence*, 4, 463-502.

McClelland, J. L. & Rumelhart, D. E. (1981). An interactive activation model of context effects in letter perception: Part I. An account of the basic findings. *Psychol. Rev.* 88, 375-407.

McClelland, J. L. & Elman, J. L. (1986). Interactive processes in speech perception: The TRACE model. In J. L. McClelland and D. E. Rumelhart (Eds.), *Parallel Distributed Processing*, 2, 58-121, Cambridge, MA: The MIT Press.

McDermott, D. V. (2001). *Mind and Mechanism*. Cambridge, MA: The MIT Press.

Mendel, J. M. & McLaren, R. W. (1970). Reinforcement learning control and pattern recognition systems, in *Adaptive Learning and Pattern Recognition Systems: Theory and Applications* (J. M. Mendel and K. S. Fu, Eds.), 287-318, New York: Academic Press.

Minsky, M. L. (1954). Theory of neural-analog reinforcement systems and its application to the brain-model problem, Ph.D. Thesis, Princeton Univ.

Minsky, M. L. & Papert, S. A. (1969). *Perceptrons*. Cambridge, MA: The MIT Press.

Minsky, M. (1979). K-lines: a theory of memory. A.I. Memo: Massachusetts Institute of Technology.

Minsky, M. (1985). *The Society of Mind*, New York: Simon & Schuster.

Moody, J. E. & Darken, C. J. (1989). Learning algorithms and networks of neurons. In *The Computing Neuron*, R. Durbin, C. Miall, and G. Michison, eds., 35-53, Reading, MA: Addison-Wesley.

Mozer, M. C. (1991). *The Perception of Multiple Objects: A Connectionist Approach*. The MIT Press.

Mozer, M. C. & Sitton, M. (1998). Computational modeling of spatial attention. In H. Pashler (Ed.). *Attention*, 341-393. London: Psychology Press.

Murata, N., Ikeda, S., & Ziche, A. (2001). An approach to blind source separation based on temporal structure of speech signals. *Neurocomputing*, 41-1(4), 1-24.

Nadaraya, E. A. (1964). On estimating regression. *Theory Probab. Applic.*, 10, 186-190.

Nakajima, T., Suzuki, T., Ohmura, H., Ishizaki, S., & Tanaka, K. (1978). Estimation of vocal tract area function by adaptive deconvolution and adaptive speech analysis system. *J. Acoustical Soc. of Japan* 34-3, 157 (in Japanese).

Nakano, K., Iinuma, K., *et al.* (1989). *Neurocomputer.* Gijutsu-Hyoron, Publishing Co. Ltd.

Newell, A. & Simon, H. (1997). Computer science as empirical inquiry: symbols and search. In J. Haugeland (Ed.), *Mind Design II*, Cambridge, MA: The MIT Press, (pp.81-110).

Nguyen Thi, H. L. & Jutten, C. (1995). Blind source separation for convolutive mixtures. *Signal Processing*, 45-2, 209-229.

Nosofsky, R. M. (1986). Attention, similarity and the identification-categorization relationship. *J. Experimental Psychology: General*, 115, 39-57.

Oatley, K. & Jenkins, J. M. (1996). *Understanding Emotions.* Cambridge, MA: Blackwell.

Oja, E. (1983). *Subspace Methods of Pattern Recognition.* England: Research Studies Press (Japanese translation: Sangyo-Tosho, 1986).

Oppenheim, A. V. & Schafer, R. W. (1975). *Digital Signal Processing.* London: Prentice Hall Int.

Orr, M. J. L. (1996). *Introduction to radial basis function networks.* [Online] Available www.cns.ed.ac.uk.

Osaka, N. (1997). *Nou-To-Ishiki (Brain and Consciousness).* Asakura-Shoten (in Japanese).

Osgood, C. E., Suci, G. J., & Tannengaum, P. H. (1957). *The Measurement of Meaning.* Urbana, IL: Univ. Illinois Press.

Papez, J. W. (1937). A proposed mechanism of emotion. *Arch. Neurol. Psychiatry*, 79, 217-224.

Parker, D. B. (1985). Learning logic: Technical Report TR-47, Center for Computational Research in Economics and Management Science, Massachusetts Institute of Technology, Cambridge, Mass.

Parzen, E. (1962). On estimation of a probability density function and mode. *Annals of Mathematical Statistics*, 33(3), 1065-1076.

Perrett, D. I., Rolls, E. T., & Caan, W. (1982). Visual neurons responsive to faces in the monkey temporal cortex. *Experimental Brain Research*, 47, 329-342.

Pfeifer, R. & Scheier, C. (2000). *Understanding Intelligence.* Cambridge, MA: The MIT Press.

Phaf, H. R., Van Der Heijden, A., & Hudson, P. (1990). SLAM: A connectionist model for attention in visual selection tasks. *Cognitive Psychology*, 22, 273-341.

Pinker, S. (1997). *How the Mind Works.* New York: W. W. Norton & Company.

Platt, J. (1991). A resource-allocating network for function interpolation. *Neural Computation*, 3-2, 213-225.

Poggio, T. & Edelman, S. (1990). A network that learns to recognize three-dimensional objects. *Nature*, 343-18, 263-266.

Poggio, T. & Girosi, F. (1990). Networks for approximation and learning. *Proc. of IEEE*, 78, 1481-1497.

Polikar, R., Udpa, L., Udpa, S. S., & Honavar, V. (2001). Learn++: an incremental learning algorithm for supervised neural networks. *IEEE Trans. Systems, Man, and Cybernetics – Part C: Applications and Reviews*, 31-4, 497-508.

Proakis, J. G. & Manolakis, D. G. (1992). *Digital Signal Processing: Principles, Algorithms, and Applications*. 2nd. Ed. New York: Macmillan.

Rabiner, L. R. & Juang, B.-H. (1993). *Fundamentals of Speech Recognition*. Prentice Hall, Inc.

Renals, S. (1989). Radial basis function network for speech pattern classification. *Electronics Letters*, 25, 437-439.

Ritter, H., Martinetz, T., & Schulten, K. (1992). *Neural Computation and Self-Organizing Maps – An Introduction*. Reading: Addison Wesley.

Rolls, E. T. (1999). *The Brain and Emotion*. New York: Oxford Univ. Press.

Rosenblatt, M. (1956). Remarks on some non-parametric estimates of a density. *Ann. Math. Stat.*, 27, 832-7.

Rosenblatt, F. (1958). The perceptron: a probabilistic model for information storage and organization in the brain. *Psychological Review*, 65, 386-408.

Rosenblatt, F. (1962). *Principles of neurodynamics*. Washington, DC:Spartan Books. * Rosenbleuth, A., Wiener, N., Pitts, W. & Garcia Ramos, J. (1949). A statistical analysis of synaptic excitation. *J. Cellular and Comparative Physiology*, 34, 173-205.

Roy, A. (2000). Artificial neural networks – a science in trouble. *SIGKDD Explorations*, 1(2), 33-38.

Rumelhart, D.E. Hinton, G. E., & Williams, R. J. (1986). Learning internal representations by error propagation. In D. E. Rumelhart and J. L. McClelland (Eds.), *Parallel Distributed Processing: Explorations in the Microstructure of Cognition*, 1, Chap. 8, Cambridge, MA: The MIT Press.

Rumelhart, D. E. & Zisper, D. (1985). Feature discovery by competitive learning. *Cognitive Science*, 9, 75-112.

Russel, J. A. (1979). Affective space is bipolar. *J. Pers. Soc. Psychol.* 37, 345-356.

Russel, S. & Norvig, P. (2003). *Artificial Intelligence: A Modern Approach*, 2nd Ed. London: Prentice Hall.

Rutkowski, T., Cichocki, A., & Barros, A. K. (2000). Speech extraction from interferences in real environment using bank of filters and blind source separation. *Proc. of Workshop on Signal Processing Applications*.

Sadasivan, P. & Dutt, D. N. (1996). SVD based technique for noise reduction in electroencephalographic signals. *Signal Processing*, 55-2, 179-189.

Sagi, B., Nemat-Nasser, C. S., Kerr, R., Downing, R. H. C., & Hecht-Nielsen, R. (2001). A biologically motivated solution to the cocktail party problem. *Neural Computation*, 13-7, 1575-1602.

Sakai, K. (2002). *Gengo-No-Nou-kagaku (Language in Brain Science)*. Tokyo: Chu-Ko, Co. Ltd. (in Japanese).

Samuel, A. L. (1959). Some studies in machine learning using the game of checkers. *IBM J. Res. Dev.*, 3:210-229. Reprinted in (1963). *Computers and Thought* (E. A. Feigenbaum and J. Feldman, Eds.), 406-450, New York: McGraw-Hill.

Sarle, W. S. (2001). [Online] *comp.ai.neural-nets FAQ*. Part 2 of 7: Learning.

Schachter, S. & Singer, J. (1962). Cognitive, social and physiological determinants of emotional state. *Psychological Review*, 69, 379-399.

Searle, J. (1992). *The Rediscovery of Mind*. Cambridge, MA: The MIT Press.

Shibata, M. (2001). *Robot-No-Kokoro, Nanatsu-No-Tetsugaku-Monogatari (The Mind of Robots: Seven Philosophical Stories)*. Tokyo: Koudan-Sha, Co. Ltd. (in Japanese).

Shigematsu, Y., Ichikawa, M., & Matsumoto, G. (1996). Reconstitution studies on brain computing with the neural network engineering. In *Perception, Memory and Emotion: Frontiers in Neuroscience*, eds. T. Ono, B. L. McNaughton, S. Molotchnikoff, E. T. Rolls, and H. Nishijo, Elsevier, 581-599.

Shimojo, S. (1999). *Ishiki-Toha-Nandaroh-Ka? (What Is Consciousness?)*. Koudan-Sha, Publishing, Co. Ltd.

Simon, H. A. (1996). *The Sciences of the Artificial*. Cambridge, MA: The MIT Press. (Japanese translation: Tokyo: Tuttle-Mori Agency, Inc.)

Smith, E. E. & Jonides, J. (1997). Working memory: a view from neuroimaging. *Cognitive Psychology*, 33, 5-42.

Specht, D. F. (1988). Probabilistic neural networks for classification mapping, or associative memory. *Proc. Int. Conf. on Neural Networks*, 1, 525-532.

Specht, D. F. (1991). A generalized regression neural network. *IEEE Trans. Neural Networks*, 2-6, 568-576.

Specht, D. F. (1990). Probabilistic neural networks. *Neural Networks*, 3, 109-118.

Squire, L. R. (1987). *Memory and Brain*. New York: Oxford Univ. Press.

Stent, G. S. (1973). A physiological mechanism for Hebb's postulate of learning. *Proc. of National Academy of Sciences of the U.S.A.*, 70, 997-1001.

Stork, D. G. (1989). Is backpropagation biologically plausible? *Proc. of Int. Joint Conf. Neural Networks*, II, 241-246, Washington, D.C.

Swanson, L. W. (1983). The hippocampus and the concept of the limbic system. In W. Seifert (Ed.) *Neurobiology for the Hippocampus*, 3-19, London: Academic Press.

Takagi, T., Fujita, M., Hasegawa, R., Shimomura, H. Yokono, J., Costa, G., & Di Profio, U. (2001). Behavior control architecture of small humanoid robot for entertainment application. *Proc. of 11th Sony Research Forum*, 100-101 (in Japanese).

Taylor, J. R. (1995). *Linguistic Categorization: Prototypes in Linguistic Theory*, 2nd Edition, Oxford Univ. Press.

Taylor, J. G., Horwitz, B., Shah, N. J., Fellenz, W. A., Mueller-Gaertner, H.-W., & Krause, J. B. (2000). Decomposing memory: functional assignments and brain traffic in paired word associate learning. *Neural Networks*, 13-8 & 9, 923-940.

Terasawa, T. (1984). *Ishiki-Ron (The Theory of Consciousness)*. Otsuki, Publishing, Co. Ltd. (in Japanese).

Theogarajan, L. & Akers, L. A. (1996). A multi-dimensional analog gaussian radial basis circuit. presented at the *IEEE Int. Symp. Circuits and Systems*, Atlanta, GA.

Theogarajan, L. & Akers, L. A. (1997). A scalable low voltage analog gaussian radial basis circuit. *IEEE Trans. Circuits and Systems II*, 44-11, 977-979.

Tomkins, S. S. (2004). *Exploring Affect: Selected Writings of Sylvan S. Tomkins*. E. V. Demos., Ed. New York: Cambridge Univ. Press.

Torkkola, K. (1996). Blind separation of delayed sources based on information maximization. *Proc. of IEEE Int. Conf. Acoust. Speech, Signal Processing (ICASSP-96)*, 5, 3509-3512.

Tsunoda, K., Yamane, Y., Nishizaki, M., & Tanifuji, M. (2001). Complex objects are represented in macaque inferotemporal cortex by the combination of feature columns. *Nature Neuroscience*, 4-8, 832-838.

Turing, A. M. (1950). Computing machinery and intelligence. *Mind*, 59, 433-460. Reprinted in (1963). *Computers and Thought* (E. A. Feigenbaum and J. Feldman, Eds.), 11-35, New York: McGraw-Hill.

Tulving, E. (1972). Episodic and semantic memory. in *Organization of Memory*, E. Tulving & W. Donaldson (Eds.) Academic Press.

Ullman, M. T. (2001). A neurocognitive perspective on language: The declarative/procedural model. *Nature Reviews Neurosci.* 2, 717-726.

Unamuno, M. (1978). *Tragic Sense of Men and Nations – Selected Works of Miguel de Unamuno*. Princeton Univ. Press.

Vapnik, V. (1995). *The Nature of Statistical Learning Theory*, Springer Verlag.

Vetter, T., Hurlbert, A., & Poggio, T. (1995). View-based models of 3D object recognition: invariance to imaging transformations. *Cerebral Cortex*, 5-3, 261-269.

Viterbi, A. J. (1967). Error bounds for convolutional codes and an asymptotically optimal decoding algorithm. *IEEE Trans. Information Theory*, IT-13: 260-269.

von der Malsburg, C. (1973). Self-organization of orientation sensitive cells in the striate cortex. *Kybernetik*, 14, 85-100.

Warren, R. M. (1999). *Auditory Perception – A New Analysis and Synthesis*. Cambridge: Cambridge Univ. Press.

Wasserman, P. D. (1993). *Advanced Methods in Neural Computing*. In Chap. 8, Radial basis-function networks (pp.147-176). New York: Van Nostrand Reinhold.

Watson, G. S. (1964). *Smooth Regression Analysis*. Sankhy, Series A, 26, 359-372.

Waibel, A. T., Hanazawa, T., Hinton, G., Shikano, K., & Lang, K. J. (1989). Phoneme recognition using time-delay neural networks. *IEEE Trans. Acoustics, Speech, and Signal Processing*, ASSP-37, 328-339.

Weintraub, M. (1985). A theory and computational model of auditory monaural sound separation. Ph.D. dissertation, Stanford University.

Werbos, P. J. (1974). Beyond regression: new tools for prediction and analysis in the behavioral sciences. Ph.D dissertation, Harvard University.

White, J., Dickinson, T. A., Walt, D. R., & Kauer, J. S. (1998). An olfactory neuronal network for vapor recognition in an artificial nose. *Biological Cybernetics*, 78, 245-251.

Widrow, B. (1962). Generalization and information storage in networks of adaline "neurons". In M. C. Yovitz, G. T. Jacobi, & G. D. Goldstein (Eds.), *Self-Organizing Systems*, 435-461, Washington, D.C.: Sparta.

Widrow B., *et al.* (1975). Adaptive noise cancelling: principles and applications. *Proc. of IEEE*, 63, 1692-1716.

Wiener, N. (1948). *Cybernetics*. Cambridge, MA: The MIT Press.

Wilson, R. A. & Keil, F. C. (Eds.) (1999). *The MIT Encyclopedia of the Cognitive Sciences*, Cambridge, MA: The MIT Press.

Wolff, G. J., Prasad, K. V., Stork, D. G., & Hennecke, M. (1993). Lipreading by neural networks: visual preprocessing, learning and sensory integration. *Proc. of Neural Information Processing Systems (NIPS'93)*, 1027-1034.

Xie, F. & Van Compernolle, D. (1996). Speech enhancement by spectral magnitude estimation – a unifying approach. *Speech Communication*, 19-2, 89-104.

Yamadori, A, (1998). *Hito-Wa-Naze-Kotoba-Wo-Tsukaeruka (Why Can Humans Use Language?)*. Tokyo: Kodan-Sha, Co. Ltd. (in Japanese).

Yamasaki, T. & Shibata, T. (2003). Analog soft-pattern-matching classifier using float-gate MOS technology. *IEEE Trans. Neural Networks*, 14-5, 1257-1265.

Index